W0262760

THINK BIG,
START SMALL

IMPRESSUM *IMPRINT*

ISBN Buch Book: 978-3-662-54981-0, ISBN eBook: 978-3-662-54997-1

Springer Vieweg
© Der/die Herausgeber bzw. der/die Autor(en) 2017
1. Auflage 2017 1st edition 2017

Alle Rechte vorbehalten All rights reserved

Bilbliografische Informationen verzeichnet die Deutsche Nationalbibliothek: www.dnb.de
The Deutsche Nationalbibliothek lists this publication in the Deutsche Nationalbibliografie: www.dnb.de

Herausgeber Publishers:
Prof. Achim Kampker, Jürgen Gerdes, Prof. Günther Schuh
Text und Redaktion Text and editing:
Holger Schaeben, www.schaebenschreibt.de
Fachberatung Specialist consulting:
Dr. Christoph Deutskens
Gestaltung, Satz und Grafiken Design, typesetting and graphics:
Christoph Niermann, www.sehwerk.de
Übersetzung Translation:
Alexandra Wahl, www.lingo-tech.com
Druck Press:
Springer

Bildnachweise Photo credits
Seite Page: 14 Wikimedia Commons; 33 InnoZ GmbH; Umschlag Cover, 35, 65, 75, 118, 167, 171, 182, 183, 185, 186, 203–206, 215, 216, 220 StreetScooter GmbH/Deutsche Post DHL; 54, 161 Fotostudio Tölle; 36, 76, 87, 89–93, 154, 172 Andreas Schmitter; 47 Wittenstein SE; 88, 190 Deutsche Post AG; 151 Kirchhoff Automotive GmbH; 168 Trumpf GmbH & Co. KG; 213 Hojabr Riahi; 215, 217 e.GO Mobile AG

Herausgeber *Publishers*
PROF. ACHIM KAMPKER
JÜRGEN GERDES
PROF. GÜNTHER SCHUH

THINK BIG,
START SMALL

STREETSCOOTER
DIE E-MOBILE ERFOLGSSTORY
Innovationsprozesse
radikal effizienter

STREETSCOOTER
THE E-MOBILE SUCCESS STORY
Radically more efficient
innovation processes

INHALT

CONTENT

C   DIE PERSPEKTIVE

C THE PERSPECTIVE

7 The StreetScooter Network Story

8 The StreetScooter Production Story

9 The StreetScooter Post Story

10 The StreetScooter Mobility Solution

11 The StreetScooter Vision

StreetScooter can be explained in four steps

DIE ENTWICKLUNG
THE DEVELOPMENT

„MANCHMAL IST ES GUT, WENN MAN EINE SCHRAUBE LOCKER HAT,

DANN HABEN
DIE

SOMETIMES IT'S GOOD TO
HAVE A SCREW LOOSE, IT GIVES
YOUR IDEAS MORE FREE PLAY.

GEDANKEN
MEHR SPIEL.

Prof. Achim Kampker
Geschäftsführer der Streetscooter GmbH
Executive Vice President E-Mobilität Deutsche Post DHL
CEO of Streetscooter GmbH
Executive Vice President E-Mobility Deutsche Post DHL

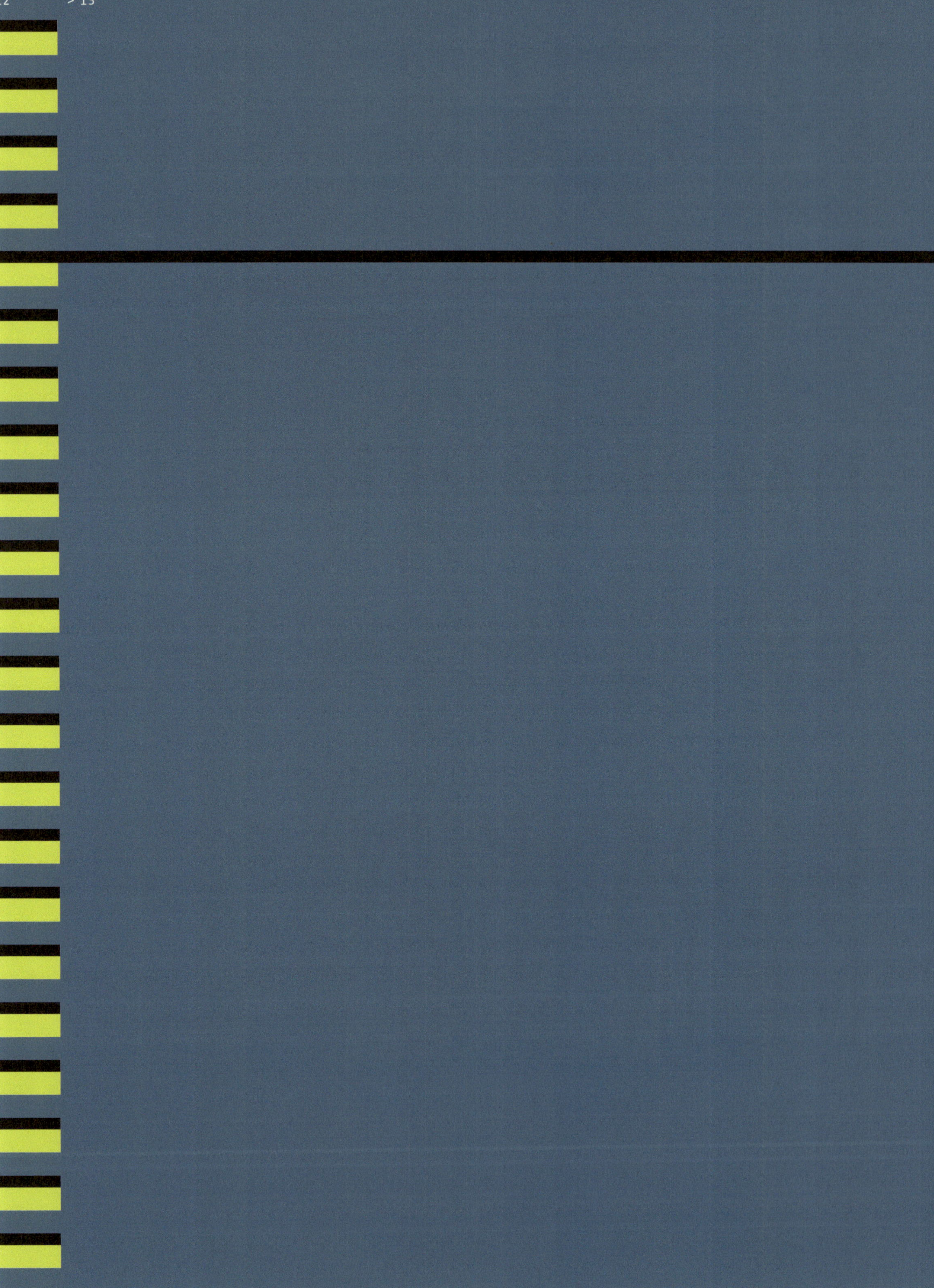

1

DREI ERFOLGSSTORIES

THREE STORIES OF SUCCESS

Geht das? Das geht! Wie man Entwicklungsprozesse durch radikales Umdenken effizienter gestaltet und so den Markteintritt schneller schafft: Drei Erfolgsstories, die zeigen, wie das geht.

Does it work? It works! How to make development processes more efficient by radically re-thinking them and thereby speeding up time to market: Three success stories that show how it works.

SCHNELLER AM START

FASTER AT THE START

Kennen Sie die Geschichte von den Amerikanern und den Russen im Weltall? Zugegeben, es gibt einige. Viele sind wahr. Manche gut erfunden. Bei allen Geschichten ging es darum, dass einer der Erste sein wollte.

Do you know the story of the Americans and the Russians in space? Granted, there are a few. Many are true. Some are pure inventions. But all the stories came down to one of them wanting to be the first.

Der erste Mensch im All. Die erste Frau im All. Der erste Weltraumspaziergänger im All. Der erste Mensch auf dem Mond. Jede dieser Geschichten erzählt auf immer andere Weise die immer gleiche Geschichte vom Wettlauf der Amerikaner und Russen im All. Die meisten Geschichten kennen wir natürlich. Aber kennen Sie auch diese hier?

The first man in space. The first woman in space. The first spacewalkers in space. The first man on the moon. Each of these stories somehow always tells the same story of the race between the Americans and Russians in space. Sure, we know most of the stories. But have you heard this one?

In dieser Geschichte geht es um einen banalen Kugelschreiber. Wir alle haben das Bild im Kopf. Es zeigt einen Astronauten in der Raumstation, wie er Einträge ins Logbuch macht oder die Ergebnisse wissenschaftlicher Bord-Experimente notiert. Schreiben? Ein einfacher Vorgang, denken Sie? Ja, vielleicht auf der Erde. Aber im Weltraum? Mit einem Kugelschreiber geht das beispielsweise da oben nicht. Der braucht die Schwerkraft. In der Schwerelosigkeit kann er nicht funktionieren, weil die Tinte nur durch die Schwerkraft nach unten zur Mine fließen kann.

Wie, fragten sich Amerikaner und Russen, würde sich das Problem lösen lassen? Erneut entbrannte zwischen ihnen ein Wettlauf im All, um den sich Geschichten ranken. Eine geht so: Die Amerikaner fühlten sich von den Russen wieder einmal provoziert, näherten sich dem Problem darum zuerst und ließen eine in ihren Augen fortschrittliche Lösung entwickeln. Das Ergebnis: 1965 schloss die NASA mit Tycam Engineering Manufacturing, Inc. für ihr Gemini-Programm einen Festpreis-Vertrag über die Lieferung von 34 mechanischen Bleistiften zum Gesamtpreis von 4.382,50 US-Dollar (128,89 US-Dollar je Stück). Tests auf der Erde, unter den Bedingungen der Schwerelosigkeit, zeigten: Der Stift fuktionierte perfekt. Ein Problem stand jedoch im Raum: Die Kosten für den Tycam-Pen waren astronomisch. Ein solcher Preis wurde in der amerikanischen Öffentlichkeit als völlig überzogen empfunden. Prompt machte die NASA den Vertrag rückgängig.

This story is about a simple pen. We all have the image in our heads. It shows an astronaut in the space station as he's making entries in the logbook or recording the results of onboard scientific experiments. Writing? A simple process, right? Yes, maybe on earth. But in space? Up there, writing with a pen doesn't work. It takes gravity.

In zero gravity, it can't work because it takes gravity to get the ink to flow downwards to the pen's reservoir.

How, wondered the Americans and Russians, could the problem be solved? A new space race broke out between them with its respective narratives growing around it. One goes like this: The Americans felt provoked by the Russians once again, approached the problem first and were able to develop what seemed to them an advanced solution. The result: In 1965 for its Gemini program, NASA signed a fixed-price contract with Tycam Engineering Manufacturing, Inc. for the supply of 34 mechanical pens for a total price of $4,382.50 ($128.89 per piece). Tests under the conditions of weightlessness on earth proved it: the pen worked perfectly. However, there was just one problem: The cost of the Tycam pen was astronomical. The American public would see such a price as utterly exorbitant. NASA quickly cancelled the contract.

Two years later, according to another story, there was a renewed attempt to develop a space pen. Supposedly in 1967 NASA itself invested over $1 million in the development of a space pen.

Zwei Jahre später, so eine andere Geschichte, folgte ein erneuter Versuch, ein Weltraumschreibgerät zu entwickeln. 1967 steckte die NASA selbst über 1 Mio. Dollar in die Entwicklung eines Weltraumstiftes. Die tüchtigsten NASA-Ingenieure machten sich daran, das Problem mit der Schwerkraft und der Tinte zu lösen und erfanden eine Art Kugelschreiber, äußerst kompliziert, der aber auch im Weltall bestens funktionierte. Dieser Stift ist in die Geschichte der Raumfahrt eingegangen.

Der Space Pen oder Astronautenstift, der übrigens nicht von der NASA, sondern von der Firma Fisher Space Pen Co. entwickelt wurde, flog zum ersten Mal 1968 mit ins All. Zur Legende wurde der Space Pen durch die fehlerhafte Berichterstattung der ARD zur Mondlandung im Jahre 1969, in der einer der Moderatoren im Studio auch die 1-Mio.-Dollar-Entwicklungsgeschichte des Bullet-Pen, wie er offiziell hieß, zum Besten gab.

Teuer hin oder her. Die Amerikaner waren trotzdem unheimlich stolz. Einerseits, weil sie vor den Russen auf dem Mond gelandet waren. Andererseits, weil ihnen die – wenn auch sehr aufwendige – Entwicklung eines Stiftes gelungen war, mit dem sie problemlos in der Schwerelosigkeit schreiben konnten. Und was war in der Zwischenzeit bei den Russen passiert? Sie hatten zwar das Rennen um die Landung auf dem Mond verloren, aber bei der „Erfindung" eines intergalaktischen Schreibgerätes hatten sie die Amis heimlich abgehängt. Viel schneller als die Amerikaner hatten die Russen eine Lösung gefunden, wie man im All schreiben

They needed to solve the problem with the gravity and the ink and so they invented a kind of pen, albeit an extremely complicated one, that would work perfectly in space. And it's this pen that's gone down in the history of space travel.

The space pen or astronaut pen, which incidentally was not developed by NASA, but by Fisher Space Pen Co., flew into space for the first time in 1968. The space pen became legendary thanks to the erroneous reporting by the ARD on the moon landing in 1969, in which one of the moderators in the studio went on about the story of the $1 million spent in developing the bullet pen, as it was officially called.

Expensive or not. The Americans were still very proud. On the one hand, because they had landed on the moon before of the Russians. And on the other hand, because they had succeeded in developing a pen – albeit at great expense – that could easily write in zero gravity. And what in the meantime were the Russians up to? They had lost the race for landing on the moon, but in the "invention" of an intergalactic writing instrument they had secretly left the Yanks far behind. Much faster than the Americans, the Russians had found a solution to how to write in space. Instead of chancing lengthy and costly development cycles, they instead presented a working solution with which they were much faster at the start: a pencil. It may not have been perfect, because the tip could break off, but it was many times cheaper than the spacey Yank pen.

konnte. Statt wie üblich langwierige und kosten-intensive Entwicklungszeiten in Kauf zu nehmen, präsentierten sie lieber eine funktionierende Lösung, mit der sie viel schneller am Start waren: den Bleistift. Der war vielleicht nicht perfekt, weil er abbrechen konnte, dafür war er um ein Vielfaches preiswerter als der spacige Ami-Stift.

Unabhängig davon, ob sich die Kugelschreiber-Bleistift-Geschichte wirklich so zugetragen hat, steht sie doch als Beispiel für ein radikal anderes Denken. Ihre Botschaft steckt vielleicht in diesem einen Satz: Die Russen konnten mit dem Bleistift sofort etwas anfangen.

Regardless of whether the pen and pencil story really happened this way, it can still be read as an example of a radically different way of thinking. Their message might be put in this one sentence: With the pencil, the Russians had a head start in making things happen.

1969

Die Amerikaner landen als Erste auf dem Mond. Die Russen denken erst mal in eine völlig andere Richtung und landen einen Coup: Bleistift vor Space Pen.

In 1969 the Americans were the first to land on the moon. Then the Russians thought in a completely different direction and pulled off a big success: pencil before space pen.

PLUG AND PLAY

Songwriter, Tonstudio, Produzent, Plattenfirma, Marketing-Millionen – vergiss es! Dieser Weg ist viel zu weit und vor allem zu teuer.

Songwriter, recording studio, producer, record company, marketing millions – forget it! This way is' much too long and above all too expensive.

Justin Bieber, Andy McKee, Ryan Higa, Shane Dawson, Fred oder Ray William Johnson – um nur einige zu nennen –, sie alle haben eines gemeinsam: Sie haben es von der Straße oder aus dem heimischen Kinderzimmer auf direktem Wege in die internationalen Charts geschafft; bis in die Top 10 oder ganz nach oben. Auf jeden Fall ganz ohne Umweg.

Was sie verbindet? Keiner von ihnen hatte zuvor einen Songwriter beauftragt, ein Tonstudio gemietet, einen Produzenten gesucht, eine Plattenfirma interessiert und schon gar keine Marketing-Millionen investiert.

Justin Bieber, Andy McKee, Ryan Higa, Shane Dawson, Fred or Ray William Johnson – to name just a few – all have one thing in common: they have made it from the street or from their parent's bedrooms directly into the international charts; skyrocketing into the top 10 or even straight to the top. Without making a single detour.

What unites them? None of them had ever commissioned a songwriter, rented a recording studio, sought a producer, got a record company interested in them or invested millions of dollars in marketing.

Stattdessen griffen sie zu ihrem mehr oder weniger guten Instrument, verließen sich auf ihre Stimme, setzten sich vor eine Videokamera und spielten oder sangen los.

Alles war unperfekt: das Bild, der Ton, die Technik, ihr ganzer Auftritt. Aber das war ihnen egal. Sie wollten, dass man hört und sieht und erkennt, was in ihnen steckt. Ihre Begabung war ihr einziges Kapital. Sie wählten den kürzesten, schnellsten und billigsten Weg in den Markt. Ihr Mittel zum Zweck: YouTube.

Da ist zum Beispiel die Geschichte von Andy McKee. Zu seinem 13. Geburtstag bekam er von seinem Vater seine erste Gitarre geschenkt. Zuhause nahm er kleine Gitarren-Videos von sich auf. Die YouTube-Trailer seiner virtuosen Performances als Fingerstyle-Gitarrist wurden bis zu 10 Millionen Mal angesehen und machten ihn innerhalb kurzer Zeit bekannt. Schließlich nahm ihn das Label Candy Rat Records unter Vertrag.

Eine andere Geschichte ist die von Justin Bieber. Man muss kein Fan seiner Musik sein, aber man muss anerkennen, dass sich seine Welt-Karriere ohne irgendwelche technischen oder finanziellen Mittel ziemlich rasant entwickelte.

Der Anfang dieser Entwicklung verlief noch ähnlich wie bei Andy McKee. Justin Drew Bieber wollte Musiker werden. Er begann mit Coverversionen bekannter Songs. Seine Mutter war es, die Videos davon machte und auf YouTube

Instead, they resorted to their more or less good instrument, relied on their voice, sat down in front of a video camera and played or sang.

Everything was imperfect: the image, the sound, the technique, their whole appearance. But they didn't care. They wanted to be heard and seen and knew what their gift was. Their talent was their only capital. They chose the shortest, fastest and cheapest way to the market. Their means to an end: YouTube.

For example, there's the story of Andy McKee. For his 13th birthday his father gave him his first guitar. He made little guitar videos of himself at home. The YouTube trailer of his virtuoso performances as a fingerstyle guitarist were seen up to 10 million times, and made him famous practically overnight. In the end, he was signed by the label Candy Rat Records.

Another story is that of Justin Bieber. You don't have to be a fan of his music, but you do have to acknowledge that he developed his world career pretty rapidly, without any technical or financial resources.

The beginning of this development was still similar to Andy McKee's. Justin Drew Bieber wanted to be a musician. He began with cover versions of popular songs. His mother was the one who made videos of them and posted them on YouTube. In 2008, the American music manager Scooter Braun became aware of Bieber's YouTube videos and Bieber got his

veröffentlichte. 2008 wurde der amerikanische Musikmanager Scooter Braun auf Biebers YouTube-Videos aufmerksam und Bieber bekam seinen ersten Plattenvertrag. Heute zählt er zu den Top Acts der internationalen Musikbranche.

Das Beispiel von Justin Bieber und den anderen genannten und ungenannten Künstlern, die es weit gebracht haben, weil sie zunächst in kurzen Schritten vorgegangen waren, zeigt, dass unkonventionelles Denken zu einem deutlich früheren Zeitpunkt handlungsfähig machen kann. Wahrscheinlich verlangte dieses Vorgehen von allen eine gewisse Naivität. Ganz sicher aber brauchte es den Glauben, dass man damit schneller vorankommen kann.

first record deal. Today he is one of the international music industry's top acts.

The example of Justin Bieber and the other named and unnamed artists who have come a long way by starting with little steps shows that unconventional thinking can make it possible to act at a significantly earlier time. It's true. This approach may have taken a certain naivety. But it definitely took the belief that this is the way to get things moving more quickly.

DRITTE ERFOLGSSTORY
THIRD STORY OF SUCCESS

EINFACH

LET'S GO KIDS!

LOSLEGEN, KINDER!

Was ist effektiver? Einfach loslegen oder bis ins kleinste Detail planen? Sie glauben, der Fall wäre klar? Lesen Sie weiter.

What is more effective? Just striking out or planning down to the littlest detail? You think this one's a no brainer? Keep on reading.

Richtig ist, dass die strategische Planung lange als eine Art Garant für sehr gute Resultate galt. Dann aber kam es zu einem Experiment mit Spaghetti, Marshmallows und einem verblüffenden Ergebnis. Teilnehmer der sogenannten Marshmallow-Challenge waren unter anderem Absolventen von Business-Schools. Gegen sie ließ man vierjährige Kindergartenkinder antreten. Das Team der Jungakademiker war schon siegessicher. Doch es kam anders: Jedes Team erhielt eine Packung ungekochter Spaghetti, einen Bindfaden, Klebeband und ein Marshmallow. Die Aufgabe bestand darin, in einem definierten Zeitrahmen von genau 18 Minuten aus den Spaghetti einen möglichst hohen Turm zu bauen, auf dessen Spitze ein Marshmallow zu platzieren war. Während die kleinen Macher einfach mit dem Bauen loslegten, dachten die großen Strategen zuerst einmal gründlich vor und zurück, erst dann fingen sie an, auf die, wie sie glaubten, vermeintlich einzig richtige Lösung hinzubauen.

Welche Gruppe hat wohl am besten abgeschnitten? Wir haben die Ergebnisse zusammengefasst: Nur 25 cm ist der Turm bei Business-School-Absolventen hoch. Kindergartenkinder bauen dagegen Türme bis zu 75 cm Höhe. CEOs konstruieren im Schnitt 60-Zentimeter-Bauwerke. Wenn nur ein einzelner CEO dabei ist, der mit seinen Sekretärinnen zusammenarbeitet, kommen 80 cm zustande.

It's true that strategic planning has long been regarded as a kind of guarantee for excellent results. But then, there was an experiment with spaghetti, marshmallows and an astonishing result. Participants of the so-called Marshmallow Challenge included recent business school graduates. Competing against them were four-year old kindergartners. The team of young academics was confident of victory. But things turned out very differently. Each team received a package of uncooked spaghetti, a ball of string, tape and a marshmallow. The task was to build the highest possible spaghetti tower, placing a marshmallow on top and all within a fixed timeframe of exactly 18 minutes. While the little builders got right down to building, the big strategists thoroughly thought things through, going back and forth, only then starting to build what they believed to supposedly be the only correct solution.

Which group clearly came out on top? We have summarized the results: The tower by the business school graduates was only 25 cm tall (just under 10 inches). But the kindergarten students built towers up to 75 cm tall (nearly 30 inches). CEOs on average build 60-centimeter structures (approx. 24 inches). When there's only a single CEO, working together with his secretaries, that average comes to 80 cm (approx. 30 inches).

Warum einfach ausprobieren effektiver ist

„Die Erklärung für diese Ergebnisse ist so einfach wie überraschend: Business-School-Absolventen halten sich nicht nur in Eitelkeiten und Rivalitätskämpfen auf. Sie sind auch darauf trainiert, strategisch auf die einzig richtige Lösung hinzubauen. Wenn das dann nicht klappt, haben sie keine Zeit mehr, etwas Neues zu bauen. Den CEOs merkt man die praktische Organisationserfahrung an. Wenn sie aber mit ihren Sekretärinnen zusammenarbeiten und nur einer den Ton angibt, werden sie eben noch besser.

Dass die Kindergartenkinder so gut abschneiden, liegt nicht nur daran, dass sie fern von Rivalitätskämpfen gut zusammenarbeiten; nein, sie sind auch einfach unbefangen genug, die Dinge einfach mal auszuprobieren. Wenn der Turm dann unterwegs zusammenkracht, versuchen sie es einfach nochmal. Mit dieser Learning-by-doing-Methode sind sie offensichtlich erfolgreicher als Business-School-Absolventen und die meisten Firmenchefs. Man sieht also, dass strategische Planung nicht immer zu den besseren Ergebnissen führen muss. Manchmal ist es besser, die Dinge einfach mal auszuprobieren."[1]

Why just trying is more effective

"The explanation for these results is as simple as it is surprising: Business school graduates are not only held back by vanity and rivalries. They are also fully trained to strategically approach the building of the single correct solution. Then if that doesn't work, they don't have any more time to build something new. With the CEOs, it's their practical organization experience you see. But if they work together with their secretaries with only one of them setting the tone, they are even better.

The reason why the kindergarteners perform so well isn't just that they work well together, far removed from rivalries. It's also because they are also simply uninhibited enough to just go ahead and try things out. If the tower collapses as they're building it, they just try building it again. By using this learning-by-doing method they are obviously more successful than business school graduates, and most CEOs. So you can see that strategic planning does not always lead to better results. Sometimes it's better to just give things a try."[1]

1 Quelle: www.imgriff.com, 03.03.2014, Die Marshmallow-Challenge
Source: www.imgriff.com, 3/3/2014, The Marshmallow Challenge

2006

Teambuilding in 18 Minuten: Business-School-Absolventen, CEOs und Kindergartenkinder treten gegeneinander an. Wer baut den höchsten Turm? Ihr Material: Spaghetti, Bindfaden, Klebeband und ein Marshmallow.

Teambuilding in 18 minutes: Business school graduates, CEOs and kindergarten kids compete against each other. Who can build the highest tower? Their materials: Spaghetti, string, tape and a marshmallow.

NICHT DAS FEHLERMACHEN BEHINDERT DEN FORTSCHRITT, SONDERN DAS FESTHALTEN AM VERDACHT,

DAS FEHLERMACHEN KÖNNTE DEN FORTSCHRITT BEHINDERN.

Prof. Achim Kampker
Geschäftsführer der Streetscooter GmbH
Executive Vice President E-Mobilität Deutsche Post DHL
CEO of Streetscooter GmbH
Executive Vice President E-Mobility Deutsche Post DHL

DIE STREETSCOOTER-GRÜNDER-STORY

THE STREETSCOOTER FOUNDER STORY

Am Anfang war das Wort und das Wort hieß Innovation. Diese Innovation war zunächst gar nicht als Fahrzeug gedacht. Die Produktionsforscher der RWTH Aachen wollten ein innovatives Produkt entwickeln, das man in einem Hochlohnland wie Deutschland weltweit wettbewerbsfähig produzieren kann. Sie ließen erst Gedanken und dann Strom fließen.

In the beginning was the word and the word was innovation. This innovation was initially not intended as a vehicle at all. The production researchers at the RWTH Aachen wanted to develop an innovative product that could be produced in a high-wage country like Germany and still be globally competitive. First they let their ideas flow and then the electricity.

ELEKTRISIERT

VON EINER IDEE

ELECTRIFIED BY AN IDEA

Ganz Deutschland tut sich schwer mit Elektroautos. Ganz Deutschland? Nein! Eine Forschungsinitiative der RWTH Aachen zeigt, dass zündende Ideen nicht nur im Silicon Valley entstehen, sondern auch an einer Hochschule tief im Westen Deutschlands.

All of Germany is struggling with electric cars. All of Germany? No, a research initiative of the RWTH Aachen shows that great ideas don't just come from Silicon Valley; they also come from a university in the heartlands of Germany's West.

Wie konnte es dazu kommen, dass ein paar innovative Köpfe der Rheinisch-Westfälischen Technischen Hochschule Aachen in ein Zukunftsprojekt einstiegen, ein Unternehmen ausgründeten und seitdem der staunenden Welt in einem kleinen, bezahlbaren und energiefreundlichen Elektrofahrzeug vorausfahren?

Betrachten wir die ersten zwei bis drei Jahre – die Gründer- und Nachgründerzeit. Rückblickend lässt sich heute sicher sagen, dass alles im Jahre 2008 im Werkzeugmaschinenlabor (kurz WZL) der RWTH Aachen begann. Die Kurve zum Elektrofahrzeug hatte man zu diesem Zeitpunkt schon genommen. Zuvor hatte die Forschungsgruppe um Prof. Günther Schuh[2] alles Mögliche an denkbaren Produkten angedacht und wieder verworfen. Ein Elektrofahrzeug war noch nicht dabei. Aber man näherte sich – mit Kopf und Bauch. Eins hatte man sich auf jeden Fall auf die Fahne geschrieben: Im Zuge des Exzellenzclusters „Integrative Produktionstechnik für Hochlohnländer" wollten die Teilnehmer der Forschungsgruppe herausfinden, welches industriell produzierte Gut, möglichst auch im großen Volumen, weltweit wettbewerbsfähig produziert werden kann, und zwar aus einem Hochlohnland wie Deutschland heraus. Nachdem man verschiedene Themenfelder abgeklopft hatte, fiel die Entscheidung auf ein Produkt, das zwar häufig in den Medien, aber leider selten auf der Straße zu sehen war: das Elektroauto. Vielleicht fand dieses Thema auch deshalb bei den Forschern großen Anklang, weil man hier eine spezielle Herausforderung sah. Was die Elektromobilität im Lande anging, ließ sich eine gewisse Ladehemmung nicht

How could it happen that a few innovators at the Rheinisch-Westfälische Technische Hochschule Aachen would join up in a future project, found a company, and thereafter to the world's astonishment take the lead driving a small, affordable and energy-friendly electric vehicle?

Let's consider the first two to three years – the founder and post founder time. Looking back today, it's safe to say that everything began in the Machine Tool Laboratory (WZL) of the RWTH Aachen in 2008. At that time, the turn to an electric vehicle had already been taken. Previously, the research group around Prof. Günther Schuh[2] had already dreamed up every conceivable product and discarded them all. An electric vehicle just wasn't an option yet. But they were getting closer – it made sense and they had a gut feeling about it. They had made an absolute commitment to one thing: As part of the Cluster of Excellence "Integrative Production Technology", the participants of the research group wanted to figure out which industrially produced goods could be competitively produced from a high-wage country such as Germany – if possible in large volume. After tossing out several topics, the decision was made for a product that was actually often seen in the media, but rarely on the road: the electric car. It's quite possible that the researchers accepted this issue with such relish since it represented a very special challenge. As far as electromobility went at the time in Germany, there was a certain undeniable resistance to uptake. The general trend wasn't really making any headway. The first electric cars were slow sellers. They were often twice

2 Professor Günther Schuh, Inhaber des Lehrstuhls für Produktionssystematik und Mitglied des Direktoriums des WZL an der RWTH Aachen und Vorsitzender der Gesellschafterversammlung der StreetScooter Research GmbH, Aachen
Professor Günther Schuh, Chair of Production Engineering and member of the Board of the WZL at the RWTH Aachen University and Chairman of the General Meeting of Shareholders of StreetScooter Research GmbH, Aachen

verleugnen. Die allgemeine Entwicklung kam nicht wirklich vom Fleck. Die ersten Elektroautos waren Ladenhüter. Häufig waren sie doppelt so teuer wie ein vergleichbares Fahrzeug mit herkömmlichem Antrieb. Die öffentliche Infrastruktur wirkte als Elektromobilitätsbremse. Heute, 2017, werden Elektroautos zwar langsam bezahlbarer und auch die Zahl der Strom-Zapfsäulen wächst stetig, aber für das alltägliche Fortkommen ist Elektromobilität bisher noch immer die Ausnahme von der Regel.

Bereits der erste StreetScooter, wie das kleine weiße Gefährt mit himmelblauen Dach auch damals schon hieß, wurde daher explizit als bezahlbares und kostengünstiges Elektroauto für den Kurzstreckenverkehr konzipiert.

Die Farben hatte man so gewählt, dass sie an das blau-weiße Logo der RWTH Aachen erinnerten. Der kleine Flitzer war das Ergebnis einer einjährigen Entwicklungs- und konzertierten Zusammenarbeit mit 80 Industriepartnern und diversen Hochschulinstituten unter der Leitung von Prof. Achim Kampker[3]. Kampker, der unter Prof. Schuh schon als Oberingenieur gearbeitet hatte und nach einer Geschäftsführertätigkeit bei einem Automobilzulieferer an die Hochschule zurückgekehrt war, widmete sich begeistert dem StreetScooter-Thema, er adoptierte dieses „E-Baby", dessen Wiege im Aachener Werkzeugmaschinenlabor in der Steinbachstraße stand, und nahm es an, väterliche Gefühle nicht ausgeschlossen. Kampker, der damals schon den Lehrstuhl für Produktionsmanagement am WZL leitete, begann ein Konsortium zum Bau des Elektrofahrzeugs ins Leben zu rufen. Und um Forschung und Entwicklung mit

as expensive as a comparable vehicle with a conventional drive. The public infrastructure acted as a brake on electromobility. Today in 2017, while electric cars are becoming more affordable, albeit slowly, and the number of electric charging stations is growing steadily, the everyday progress of electromobility has remained the exception to the rule.

Even the first StreetScooter – as the little white vehicle with the sky blue roof was called even then – was therefore explicitly designed as an affordable and cost-effective electric car for short journeys.

The colors had been chosen so that they reminded you of the blue and white RWTH Aachen logo. The little "scooter" was the result of a two-year development and concerted cooperation with 80 industrial partners and several university institutes under the direction of Prof. Achim Kampker[3]. Kampker, who had already worked under Prof. Schuh as chief engineer and after a CEO position at an automotive supplier returned to the university, devoted himself enthusiastically to the StreetScooter topic. He adopted this "e-Baby" whose cradle stood in the Aachen Machine Tool Laboratory in Steinbachstrasse and took it on, with some amount of fatherly feeling. Kampker, who was already heading the Chair of Production Management at the WZL at that time, began a consortium to launch the building of the electric vehicle. Thus, StreetScooter GmbH was soon born in order to advance research and development with partners from the business world. Under Kampker's leadership, the university and

3   Professor Achim Kampker, Leiter des Lehrstuhls für Production Engineering of E-Mobility Components (PEM) an der RWTH Aachen und CEO StreetScooter GmbH und Bereichsleiter Elektromobilität bei Deutsche Post
    Professor Achim Kampker, Head of the Department of Production Engineering of E-Mobility Components (PEM) at the RWTH, Aachen and CEO StreetScooter GmbH and Electromobility Division Manager at Deutsche Post

Prof. Dr. Andreas Knie
Geschäftsführer InnoZ GmbH, Leipzig,
Innovationszentrum für Mobilität
und gesellschaftlichen Wandel
CEO of InnoZ GmbH, Leipzig,
Innovation Center for Mobility
and Social Change

„Damit sich elektrische Fahrzeuge als zentraler Baustein einer neuen, digital vernetzten Verkehrslandschaft durchsetzen können, kann man sich nicht alleine auf die Autoindustrie verlassen. Das Projekt StreetScooter zeigt, wie kaum ein anderes Vorhaben, wie aus der akademischen Landschaft mit klugen Verbindungen zur Industrie dann wirkliche Innovationen entstehen, wenn mutige und grenzüberschreitende Schritte gegangen werden."

"For electric vehicles to prevail as a central component of a new, digitally networked traffic landscape, you can't rely solely on the auto industry. The StreetScooter project shows, as few others have, how real innovations can emerge from the academic landscape with smart links to industry when bold and transboundary steps are taken."

Partnern aus der Wirtschaft voranzutreiben, entstand schon bald die StreetScooter GmbH. Unter Kampkers Führung bewiesen Hochschule und ausgewählte Netzwerkpartner aus der Automotive-Industrie, dass es möglich ist, in nur drei Jahren ein nachhaltiges Elektroauto in die Serie zu bringen, das zudem auch noch bezahlbar war. Bezahlbarkeit war das, was von Anfang an im Fokus aller Aktivitäten stand. Es ging um die optimale Kombination aus Wirtschaftlichkeit und Ökologie. Das war die Vision. Nicht höher, schneller, weiter lautete das Motto, sondern einfacher, effizienter, günstiger.

Viele E-Fahrzeug-Start-ups sind im Premium-Segment eingestiegen. Mit dem StreetScooter sollte ein anderer Weg eingeschlagen werden. Die Idee: Im Low-Cost-Segment starten und zeigen, dass Elektromobilität auch im Massenmarkt funktionieren kann. Dabei reizte die Macher die Entwicklung einer wettbewerbsfähigen Produktion mehr, als das schnellste und leichteste Produkt auf die Straße zu bringen.

In der Konzeptphase gab es den Produktionsprototypen vor dem Fahrzeugprototypen. Das Ziel der Serienproduktion hatte man sich von Beginn an auf die Fahne geschrieben. Ein Ziel, das für viele Aktivitäten relevant war:

selected network partners from the automotive industry proved that it was possible to develop a sustainable electric car in just three years, and one that was affordable to boot. From the very start, affordability was at the center of all activities. That affordability depended on finding the best combination of economy and ecology. That was the vision. Not higher, faster, further as was the fashion; instead, it was simpler, more efficient, and cheaper.

Many e-vehicle start-ups entered the premium segment. But with the StreetScooter, another route had to be taken. The idea? To start in the low-cost segment and show that electromobility can also work in the mass market. In this case, the maker was more attracted by the development of a competitive production than by a desire to put the fastest and lightest product on the road.

In the design phase, there were production prototypes before there were vehicle prototypes. The goal of series production was something everyone had committed to right from the outset. A goal that was relevant to many activities: clear target costs at the very beginning, a selection of components with potential series suppliers,

klare Zielkosten schon am Anfang, Auswahl von Komponenten mit potenziellen Serienlieferanten oder Produktionstechnologien mit niedrigen Investitionen. Letztendlich wurde aus der Vision des Low-Cost-Fahrzeugs schnell auch der Ansatz der Low-Invest-Produktion und vor allem der Low-Cost-Entwicklung.

Premiere auf der IAA Frankfurt 2011

Nicht die Perfektion des Konzepts war der Antrieb, sondern dessen Umsetzung. Und so wurde der StreetScooter zum ersten Mal im Jahre 2011 als Modell „A12" in Frankfurt auf der Internationalen Automobilausstellung (IAA) vorgestellt. Der Kleine stieß auf großes Interesse und Prof. Kampker auf die Bundeskanzlerin. Ein Jahr zuvor, 2010, war die StreetScooter GmbH gegründet worden. Die Begegnung mit Kanzlerin Merkel hatte weitreichende Folgen. Sie reichten auf jeden Fall bis nach Bonn in den Posttower. Diverse Zeitungen hatten über das junge Unternehmen und über die Begegnung der Kanzlerin mit Prof. Kampker auf der IAA berichtet. Einen dieser Artikel bekam Jürgen Gerdes in die Hand. Das Mitglied im Vorstand der Deutschen Post DHL Group las und es dauerte nicht lange, bis er sich nach Aachen aufmachte, wo ein Start-up namens StreetScooter GmbH offensichtlich ein gleichnamiges Elektrofahrzeug entwickelt hatte, nach dem er in ähnlicher Form schon lange suchte.

Ganz auf die Bedürfnisse von City-Automobilisten zugeschnitten, war das Ur-Modell des StreetScooter individuell gestaltbar. Konzipiert

or production technologies requiring low investment. Ultimately, from the vision of a low-cost vehicle also came the approach of low-invest production and most importantly low-cost development.

Premiere at the IAA Frankfurt 2011

The impetus wasn't the perfection of the concept but its implementation. And thus, the StreetScooter was presented for the first time in 2011 as model "A12" at the International Motor Show (IAA) in Frankfurt. The little one was met with great interest and Prof. Kampker met with the German Chancellor. A year earlier, in 2010, StreetScooter GmbH had been founded. The encounter with Chancellor Merkel had far-reaching consequences. They definitely reached Deutsche Post's Post Tower headquarters in Bonn. Various newspapers had reported on the young company and about the Chancellor's meeting with Prof. Kampker at the IAA. One of these articles came across the desk of Jürgen Gerdes. The member of the board of the Deutsche Post DHL Group read it and in short order, he set out for Aachen, where a start-up called StreetScooter GmbH had obviously developed an eponymous electric vehicle, something he'd been looking for in a similar form for a long time.

As the original model of StreetScooter was entirely tailor-made to meet the needs of city motorists, it was individually customizable. Designed with two to three seats, an additional child seat could also be installed. Transporting

Heute schon Geschichte: ein Schwarz-Weiß-Blick in die junge Historie (2010) des StreetScooters
Making history: a black and white look at the early days (2010) of the StreetScooter

mit zwei bis drei Sitzen ließ sich außerdem ein zusätzlicher Kindersitz montieren. Transportieren war auch möglich. Der Kofferraum fasste leicht einen Alltagseinkauf inklusive Wasserkiste. Der StreetScooter war ein typischer Stadtflitzer, speziell gedacht für den Endverbraucher. Das war natürlich nicht ganz das, was DHL-Vorstand Gerdes suchte. Er brauchte mehr ein „Werkzeug" denn ein Kleinfahrzeug, sprich einen Transporter. Im Auftrag der DHL entwickelten die Aachener ein speziell auf den Arbeitsalltag im Zustelldienst maßgeschneidertes „Werkzeug". 2012 konnte DHL das Nutzfahrzeug „Work" präsentieren, ein Flottenfahrzeug, das nach dem Konzept des Fleet Customization[4] entstanden war. Zwei Jahre später kaufte der Konzern die Firma StreetScooter.

things besides passengers was also possible. The trunk could easily hold the everyday shopping, including a crate of water. The StreetScooter was a typical city car, specially designed for the end user. Of course, that was not quite what DHL Board Member Gerdes was after. He needed more of a "tool", a small vehicle, a van basically. On behalf of DHL, the Aachen team developed a "tool" specially tailored for the everyday work of postal delivery. In 2012, DHL presented the "Work" utility vehicle, a fleet vehicle that was created according to the concept of Fleet Customization[4]. Two years later, the group bought the StreetScooter company.

4 Fleet Customization, wirtschaftliche Individuallösungen für kundenindividuelle Fahrzeugkonzepte in kleinen und mittleren Serien
Fleet Customization, cost-efficient individual solutions for customized vehicle concepts in small and medium-sized series

„WIR HABEN GEZEIGT, WIE MAN SCHLAGARTIG DIE MARKTEINTRITTSBARRIEREN FÜR FREMDE DRITTE, ALSO QUEREINSTEIGER, SUBSTANTIELL REDUZIERT."

"WE HAVE SHOWN HOW TO ABRUPTLY AND SUBSTANTIALLY REDUCE THE MARKET ENTRY BARRIERS TO EXTERNAL THIRD PARTIES – I. E., NEWCOMERS."

Prof. Günther Schuh
Inhaber des Lehrstuhls für Produktionssystematik und Mitglied des Direktoriums des WZL an der RWTH Aachen und Vorsitzender der Gesellschafterversammlung der StreetScooter Research GmbH, Aachen
Chair of Production Engineering and Board Member of the WZL at the RWTH Aachen University and Chairman of the General Meeting of Shareholders of StreetScooter Research GmbH, Aachen

Think Big, Start Small: Haben Sie wirklich groß gedacht, als Sie an das Projekt StreetScooter herangingen?

Prof. Schuh: Wir haben einen ungewöhnlichen Prozess gestartet, wie man an eine Fahrzeugentwicklung rangeht.

Jetzt ist der StreetScooter aber doch irgendwie etwas Großes geworden, oder?

So eine Tesla-Story ist eine Think-Big-Story. Streng genommen hat Elon Musk [Gründer der Firma Tesla] auch klein angefangen, der hat einfach gesagt: Ich nehme ein fertiges Auto, also den Lotus, und operiere da über fertige Batteriezellen einen relativ wenig komplexen Motor rein. Und dann macht der ein Auto und ist selber überrascht, dass trotz seines Vorbildes die Branche quasi nicht reagiert. Aber das Big Picture, das hat er trotzdem von vornherein auf dem Schirm gehabt, das hatten wir gar nicht, weil das Think Big immer eine Finanzierungsgeschichte ist. Big und Small ist für mich gar nicht so entscheidend. Viel wichtiger ist Think und Start.

Aber am Ende des Tages steht nun doch etwas Großes da, was draus wurde. Das ist vielleicht nicht Teil der Vision gewesen. Stoßen Sie sich daran?

Ich habe nichts dagegen, man muss jetzt trotzdem einmal sagen, unter groß wird die Öffentlichkeit eher solche Dinge wie Tesla verstehen. Musk hatte ein ganz großes Bild, wo er einmal hin will: Ich will mal der führende ökologische Mobilitätsanbieter werden. So eine visionäre Musk-Story, die hat uns gar nicht umgetrieben. Der Knaller bei uns ist eigentlich ein anderer. Nämlich der: Wenn man genauer auf die Technologie, vor allem auf die mögliche Produktionstechnologie schaut, haben wir gezeigt, wie man auf einmal schlagartig die Markteintrittsbarrieren für fremde Dritte, also Quereinsteiger, substantiell reduziert.

Das heißt genauer?

Eigentlich gibt es vier unüberwindbare Hindernisse für einen Quereinsteiger in die

Think Big, Start Small: Did you really think big when you approached the StreetScooter project?

Prof. Schuh: We launched an unusual process for how to approach vehicle development.

But since then, the StreetScooter has become something of a big thing, hasn't it?

The Tesla story is a kind of Think Big story. Strictly speaking, Elon Musk [founder of Tesla] also started small, simply saying: I'll take a finished car, the Lotus say, and mount a relatively simple engine into it over finished battery cells. And then he makes a car, and to his own surprise finds that in spite of his model, the industry hardly takes any notice. But the big picture he nevertheless had in view right from the outset, we didn't have, because the Think Big story is always a story about funding. Big and Small is not so critical for me. Think and Start is much more important.

But at the end of the day, there's still something Big that came out of it. This may not have been part of the vision. Does that irk you?

I have nothing against it, but now you do need to reiterate that by Big, the public tends to understand things like Tesla. Musk had a very big picture of where he really wanted to go: I simply want to be the leading ecological mobility provider. A visionary Musk story hasn't worried us in the least. What does it for us is actually something different, and it's this: If you take a closer look at technology, especially at the possible production technologies, we have shown how to abruptly and substantially reduce the market entry barriers to external third parties – i.e., newcomers.

So, that's what you mean by a closer look?

Actually there are four insurmountable obstacles for a newcomer to the automotive industry. The barrier to entry is so great because first off, you need a brand, second, a distribution channel, third, a self-supporting body structure, and fourth, a high-tech powertrain unit. In particular, the self-supporting body structure is extremely

5 OEM, englisch Original
   Equipment Manufactu-
   rer; übersetzt Originalausrüstungshersteller.
   Hier sind darunter die
   etablierten Automobil-
   hersteller zu verstehen.
   OEM, Original Equip-
   ment Manufacturer.
   These include the
   established automo-
   tive manufacturers.

Autoindustrie. Die Markteintrittsbarriere ist so groß, weil: erstens braucht man eine Marke, zweitens einen Vertriebskanal, drittens eine selbsttragende Bodystruktur, viertens ein Hightech-Powertrain-Aggregat [Antriebsstrang]. Vor allem die selbsttragende Bodystruktur ist extrem werkzeugintensiv und kapitalintensiv in der Veredelung, in der Lackierung. Das kann technologisch ein OEM[5], er kann Karosserie und er kann Powertrain. Wenn man dann ein bisschen reinguckt, was bei einem Elektrofahrzeug relevant ist, dann ist das Schwierige die Batterie, und da gibt es zurzeit kein Dominant Design, also ist keiner der möglichen Anbieter in einer Poleposition. Keiner wird, ähnlich wie bisher beim Motor, die Batterie selber produzieren. Bei Elon Musk ist das nur im Moment noch so. Wenn man das Thema „selbsttragende Bodystruktur" mal genauer betrachtet, kommt man bei der Durchleuchtung der technologischen Möglichkeiten zu dem interessanten Schluss: Upps! Das probate Konzept „Selbsttragende Karosserie" ist bei einem Elektroauto gar nicht mehr nötig, ich kriege sogar Freiheitsgrade, ich kriege Effekte bei kleinerer Stückzahl, wenn ich es anders mache. Und ich kann sogar noch ein paar technische Vorteile daraus ziehen. Genauso war das in der entscheidenden Phase, als wir das erste StreetScooter-Fahrzeug entworfen haben.

Was war denn Ihre eigentliche Motivation?

Wir wollten im Zuge des Exzellenzclusters „Integrative Produktionstechnik" herausfinden, welche industriell produzierten Güter man – möglichst im großen Volumen – auch übermorgen noch weltweit wettbewerbsfähig aus Hochlohnländern wie Deutschland heraus produzieren kann. Wir haben uns alles Mögliche vorgenommen: Werkzeugmaschinen, Roboter, Elektronikteile. Und dann, nach etwa zwei, drei Jahren, habe ich gesagt: Vielleicht passiert noch einmal so eine ähnliche Entwicklung wie bei den Hybrid-Fahrzeugen. Das kann beim Elektrofahrzeug vielleicht auch passieren. Es gab nämlich die latente Befürchtung, dass diese Autos zu teuer würden.

tool-intensive and capital-intensive in the finishing and the coating process. Technologically, an OEM[5] can do that; they can do the chassis and the powertrain. But if you look into it a bit, what's relevant for an electric vehicle – the hard part – is the battery, and right now there's no dominant design, so none of the potential suppliers is in a pole position. Just like with engines, nobody's going to be producing their own batteries. Elon Musk is in the same situation, but only temporarily. If you take a bit of a closer look into the topic of "self-supporting body structure", as you screen through the technological possibilities, you come to an interesting conclusion: Whoops! The tried and true concept of a self-supporting body is no longer necessary in an electric car. If I do it differently, I actually gain a certain degree of freedom, I gain different effects at a smaller production quantity. And I can even derive a few technical advantages from it too. That was exactly how it was in the crucial phase, when we designed the first StreetScooter vehicle.

So, what was your original motivation?

As part of the Cluster of Excellence "Integrative Production Technology", we wanted to figure out which industrially produced goods could be competitively produced from high-wage countries such as Germany – in large volume if possible – not just now, but in the future, and internationally. We got down to work on everything possible: machine tools, robots, electronic parts, you name it. And then, after about two or three years I said: Maybe there will be another development like there was in hybrid vehicles. That could happen with electric vehicles, too. There was always the latent fear that these cars would be too expensive.

A typical application area for production researchers?

Exactly. What can I do now to get the car cheaper? So, that's the actual research question

Typisches Einsatzgebiet für Produktionsforscher?

Genau. Was kann ich jetzt tun, um das Fahrzeug günstiger zu bekommen? So ist dann die eigentliche Forschungsfrage, die zum StreetScooter geführt hat, entstanden. Nämlich: Ist es mit heutiger Technologie, und das heißt unter anderem auch mit heutiger Batterietechnologie, in der Erwartung, dass Batterien über die Zeit noch deutlich günstiger werden – möglich, ein bezahlbares, robustes Elektroauto zu bauen?

Wie haben Sie „bezahlbar" verstanden?

Bezahlbar haben wir als Total Cost of Ownership definiert, nur der Zeitraum war nicht klar, das hat sich später mit der Post geklärt. Die Frage war, ob unser Zustellfahrzeug in diesem Normalzyklus von sieben Jahren, in dem die Post denkt, dasselbe oder sogar ein bisschen weniger kostet als die Fahrzeuge, die ersetzt würden wie VW T5 Transporter und Caddies.

Anfangs hatten Sie nicht für einen Transporter geplant, sondern für einen Kleinwagen, richtig?

Das war so ein Dreieinhalbsitzer, dieser StreetScooter A12. An dem wollten wir das beweisen und haben es auch bewiesen. Nach meiner Kenntnis ist es immer noch so, dass der StreetScooter das einzige bezahlbare Elektroauto ist, weil es die Total Cost of Ownership eines Verbrenners unterbietet. Noch entscheidender mit Bezügen auf Start Small ist aber, dass es das TCO bei geringen Stückzahlen unterbietet.

Gibt es auch noch Potenzial nach unten?

In der Batterie liegt natürlich noch großes Potenzial. Man hat in der Vergangenheit irgendwann einmal den Sprung auf die Lithium-Ionen-Batterie geschafft, die den besonderen Vorteil hat, dass man sie Peakmäßig entladen kann, was mit den alten Autobatterien nicht funktioniert.

Der StreetScooter sollte flink und günstig sein.

Das war halt auch der Reiz. Wir haben ein Chassis[6] gebaut, aus Blech, das noch lackiert

to emerge that led to StreetScooter. Namely: Is it possible to build an affordable, robust electric car using today's technology – and that means, among other things using today's battery technology – in the expectation that the batteries will become significantly cheaper over time.

What was your idea of "affordable"?

We defined affordable as the total cost of ownership, just that the timespan of that ownership wasn't clear, which I later clarified with Deutsche Post. The question was whether our delivery vehicle would cost the same or even a bit less in this normal cycle of seven years, in which the Post was thinking, than the vehicles that would be replaced, such as VW T5 transporter and Caddies.

You hadn't initially planned for a transporter, but for a compact car instead, right?

That was a kind of three-and-a-half seater of this StreetScooter A12. We wanted to prove it with that model and that's just what we did. To my knowledge, the StreetScooter is still the only affordable electric car, because it undercuts the total cost of ownership of a combustible engine car. More importantly, however, with reference to Start Small, is that it undercuts the TCO for small quantities.

Is there any potential for that to drop even more?

In the battery of course, there's still great potential. At some point in the past, they succeeded in making the leap to the lithium-ion battery, which has the particular advantage that it can peak discharge, which you can't do with the old car batteries.

The StreetScooter had to be quick and cheap.

That was basically its appeal. We built a chassis[6] out of sheet metal, that still needed to be

6 Chassis, tragende
Teile von Fahrzeugen:
Fahrgestell, Rahmen,
Untergestell
Chassis, load-bearing
parts of vehicles:
Chassis, frame,
undercarriage

werden muss, aber mattschwarz mit einer einfachen KTL-Lackierung[7]. Und dann kommen eben schön aussehende, leichte Thermoform-Teile drauf. Zwei-, dreihunderttausend Euro Werkzeugkosten reichen aus, um die gesamte Außenhautteile zu formen. Anders bei einem selbsttragenden Body. Erstmal habe ich um die 100 Mio. Euro Werkzeugkosten und dann kommt noch der Zulassungsprozess auf alle Strukturbauteile. Mit dem StreetScooter muss ich zwar auch durch den Zulassungsprozess, aber nur, wenn ich Gravierendes verändere. Unterm Strich bedeutet das wesentlich weniger Aufwand.

Wie ist die Post eigentlich aufmerksam geworden auf diese Entwicklung?

Um unseren Ersten, den A12, bauen zu können, hat Professor Kampker mit einer Gruppe von Unternehmern verhandelt. Da waren verschiedene dabei, die dann teilweise auch Gesellschafter wurden wie die Firmen Kirchhoff und Wittenstein. Eigentlich wollte ich im Sinne des Exzellenzclusters nur ein Konzept abliefern, um zu beweisen, dass unser Konzept funktioniert. Zuerst hat sich keiner dafür interessiert. Weder in der Wissenschaftsszene noch in der Automobilszene. Und dann war dieses Kampker-Konsortium aber doch so dynamisch, dass sie gesagt haben, dann fangen wir eben einfach mal an. So haben wir dann binnen kürzester Zeit den A12 gebaut, haben ihn auch von Fachjournalisten testen lassen und sind dann damit auf der IAA 2011 gewesen. Und in diesen A12 hat sich Frau Merkel reingesetzt und es kam zu einem kurzen Austausch zwischen ihr und Professor Kampker. Gut, weiter so!, hat sie gesagt. Danach waren wir natürlich alle tierisch motiviert.

Da hat sie noch nicht gesagt: Das schaffen wir!

Nein, aber über diesen Messeauftritt ist die Deutsche Post auf uns aufmerksam geworden und hat gefragt: Ist das tatsächlich so ein modulares – wir sagen heute Frugal-Engineering[8]-Konzept – und könnte man daraus auch ein etwas größeres Paketzusteller-Fahrzeug machen?

painted, but matte black with a simple CDL coat[7], and then we just added nice looking light thermoform parts to it. Two to three hundred thousand euros in tool costs are enough to form the entire outer skin panels, as opposed to a self-supporting body. First, I have around 100 million Euros in tool costs, and then there is the approval process on all structural components. With the StreetScooter, I also had to go through the approval process, but then only if I drastically changed something. The bottom line is that it's significantly less effort and outlay.

How did Deutsche Post actually become aware of this development?

To build our first one, the A12, Professor Kampker negotiated with a group of entrepreneurs. The group was diverse, and some went on to become shareholders, such as the company Kirchhoff and Wittenstein. Actually, in the context of the excellence cluster, I just wanted to turn out one concept, to prove that our concept worked. At first, no one was interested in it. Neither in the scientific community nor in the automotive scene. And then this Kampker consortium was really so dynamic and said, OK, let's go ahead and do it. So, we then built the A12 in short order, even having it tested by trade journalists and then there we were with it at the IAA 2011. And into this A12, Chancellor Merkel sat down behind the wheel, and there was a brief exchange between her and Professor Kampker. Great, keep up the good work! she said. Naturally after that we were all incredibly fired up.

But that was before she said those famous words: "Das schaffen wir!" (We can do it!)

True, but it was through this trade fair that the Deutsche Post first noticed us and asked: Is that really a modular – what we'd today call a "frugal engineering"[8] – concept – and could you also make a slightly larger package delivery vehicle from it?

7   KTL, kathodische Tauchlackierung; ein elektrochemisches Verfahren, bei dem das Chassis in einem Bad aus wässrigem Tauchlack unter einer Gleichspannung von 3.000 Ampere und 220 bis 290 Volt beschichtet wird.
CDC, cathodic dip coating; an electrochemical method, in which the chassis is coated in a bath of aqueous coating material under a DC voltage of 3,000 amps and between 220-290 volts.

8   Frugal Engineering, der Begriff wurde von Carlos Ghosn geprägt, Chef von Renault und Nissan, der erklärte: „Frugal Engineering heißt, mit weniger Ressourcen mehr zu erreichen."
Frugal engineering, the term was coined by Carlos Ghosn, CEO of Renault and Nissan, who explained: "Frugal engineering is achieving more with fewer resources."

Wobei wir doch wieder ein bisschen beim Big wären.

Ja, eigentlich war das, was die Post dann skizzierte, etwas größer als die größte Version, die wir vorgesehen hatten. Wir haben gesagt, na ja, das hat nachher mit dem Compartement wahrscheinlich ein paar Überhänge, aber wir sagen jetzt mal mutig: ja. Die Post hat dann etwas investiert und wir haben dafür einen ersten Entwurf geliefert. Professor Kampker hat dazu eine Costumer Journey organisiert, eine Interaktion mit dem Kunden, indem er sich Meinungen von einer guten Hundertschaft von Zustellern in X Workshops eingeholt hat. Diese ganzen Anforderungen sind in die Entwicklung eingeflossen. Aus der Praxis für die Praxis.

Die StreetScooter ist eine Ausgründung der RWTH Aachen, jetzt ganz ein Unternehmen der Deutschen Post. Gut für Sie?

Wir waren uns eben sicher, wenn wir es an die Post verkaufen: Die Post bringt's. Wir sind ja auch ein bisschen Erfinder, wir wollten am Schluss einfach unser Auto auch massenhaft auf der Straße sehen. Das ist dann schon was Großes.

Am Schluss eine grundsätzliche Frage an den Ingenieur in Ihnen: Was verhindert eigentlich radikale Neu-Entwicklungen?

Fast alle etablierten Unternehmen diskutieren und erarbeiten sich den potenziellen Erfolg, indem sie vom bestehenden Produkt her denken und von da aus überlegen, was man an diesen Produkten noch weiterentwickeln und verbessern könnte. Das verhindert fast per se ein gewisses radikales Neudenken. Sich stattdessen in die Kunden zu versetzen – und wie die Marketingleute sagen, Customer Journeys zu unternehmen – und eigentlich viel radikaler aus der Customer Perception zu kommen, das macht uns meines Erachtens Silicon Valley vor.

Ja, wir erwarten von den Apples dieser Welt, dass sie immer wieder Neues bringen.

Aber eigentlich müssen wir uns nicht verstecken. Bei aller Bescheidenheit: Auch wir können das.

Which in a way brings us back round to Big.

Yes, actually what the Deutsche Post then sketched out was somewhat larger than the largest version we had yet planned for at that point. We said, well, there will probably be a few overhangs with the compartment later, but right now, we're boldly saying: Yes. The Deutsche Post then invested a bit and we delivered a first design. Professor Kampker arranged a customer journey for it, an interaction with the customers, where he collected opinions from a good hundred delivery staff in X workshops. All these requirements were incorporated into the development. Taken from practice to be put into practice.

The StreetScooter is a spin-off of the RWTH Aachen, now a wholly owned company of the Deutsche Post. Is that good for you?

We were just sure that if we sold it to the Deutsche Post, the Deutsche Post would really use it. Because in a way, we're also inventors, at the end of the day we simply wanted to see loads of our cars on the street. So that's already something Big.

In the end, a fundamental question for the engineer in you: What actually stands in the way of radical new developments?

Almost all established companies discuss and work out potential success by starting from the existing product on out, thinking from that starting point how they might further develop and improve the product. This prevents some radical rethinking almost per se. Instead, putting yourself in the shoes of the customer – and as the marketing people say to take customer journeys – and to actually get much more radically away from customer perception, that's something in which Silicon Valley is ahead of us on in my view.

True, from the Apples of this world we expect that they constantly come out with something new.

But to be honest, we have reason to be proud too. In all modesty: We can do that, too.

WENN MAN ERKANNT HAT, DASS ES SO NICHT **WEITERGEHEN** KANN, MUSS MAN EBEN

WHEN YOU RECOGNIZE THAT THINGS CAN'T GO ON THIS WAY, YOU HAVE TO THINK FARTHER TO GET FARTHER.

WEITER DENKEN, UM WEITER ZU KOMMEN.

Prof. Achim Kampker
Geschäftsführer der Streetscooter GmbH
Executive Vice President E-Mobilität Deutsche Post DHL
CEO of Streetscooter GmbH
Executive Vice President E-Mobility Deutsche Post DHL

3

DIE STREETSCOOTER-MEILENSTEINE 2008 – 2017

THE STREETSCOOTER MILESTONES 2008 – 2017

Von Null auf „A12" in einem Jahr. Vom Beginn der Post-Ära bis heute. Was in einem überschaubaren Zeitraum von ein paar Jahren aus dem StreetScooter geworden ist – hier ein Schnelldurchlauf.

From zero to "A12" in one year. From the beginning of the Post era to today. What has become of the StreetScooter in a reasonable period of a few years – here's a quick rundown.

INNOVATION

TO GO!

Der Projekt-Kick-off liegt nur wenige Jahre zurück. 2008 war es, als der Startschuss fiel. Ort der Handlung: das WZL (Werkzeugmaschinenlabor) der RWTH Aachen. Das Kind wurde schon bald auf den Namen StreetScooter getauft und eine E-mobile Erfolgsstory nahm ihren Lauf. 2009 war die Stunde der ersten Vorstudie gekommen. Die Vorstudie war ein Konzept aus CAD-Modell plus Excel-Berechnungen plus Simulationen plus Powerpoint. Dann wurde eine alte Industriehalle angemietet und die Produktion und das Fahrzeug im Cardboard Engineering (aus Pappe) 1:1 aufgebaut. Konzeptionell war das der erste echte Schritt aus der Theorie heraus – von zwei- zu dreidimensional.

2010 folgte die formale Gründung der StreetScooter GmbH als universitäre Ausgründung.

The project kick-off was just a few years ago. It was in 2008 when the starting signal sounded. Scene: The WZL (Laboratory for Machine Tools) of the RWTH, Aachen. The child was soon thereafter christened StreetScooter and an e-mobile success story began its way in the world. 2009 saw the first preliminary study. The preliminary study was a concept made from CAD model plus Excel calculations plus simulations and PowerPoint. Then an old industrial hall was rented and the production and the vehicle built in Cardboard Engineering on a 1:1 scale. Conceptually, this was the first real step away from the theory – from two to three-dimensional.

In 2010 followed the formal establishment of StreetScooter GmbH as a university spin-off.

Dr.-Ing. Bernd Schimpf
Vorstand WITTENSTEIN SE
Member of the Board WITTENSTEIN SE

„Das StreetScooter-Projekt war für alle Beteiligten ein voller Erfolg. Nicht nur dank der exzellenten Projektleitung, auch aufgrund der außerordentlich kompetenten Besetzung der einzelnen Themenfelder war eine sehr effiziente und effektive Entwicklung möglich. Zudem konnten radikale Ansätze schnell und ohne allzu viele Restriktionen umgesetzt werden, da es sich um ein komplett neues Fahrzeugkonzept handelte. So machen Innovationen Spaß!"

"The StreetScooter project was a great success for everybody involved. Not only thanks to the excellent project management, but also to the extremely competent experts from the various subject fields, which made the development extremely efficient and effective. In addition, radical approaches could be quickly implemented and without too many restrictions, because it was a completely new vehicle concept. That's what makes innovations so much fun!"

Gleichzeitig dehnte man den Aufbau der Netzwerkstruktur, bestehend aus zeitweise rund 80 Industriepartnern, aus. Die Automotive-Partnerunternehmen hatten großen Anteil an der Entwicklung des StreetScooters. Ohne diese Integrationsstrategie wäre der Erfolg des Projektes nicht möglich gewesen. Einen Namen machte sich der StreetScooter dann ein Jahr später, als er seinen großen Auftritt vor internationalem Fachpublikum hatte. Das Ur-Modell „A12" (der erste funktionsfähige Primotyp) wurde auf der IAA 2011 in Frankfurt präsentiert. Als durch diesen ersten öffentlichen Auftritt die Deutsche Post auf den StreetScooter aufmerksam wurde und einen Entwicklungsauftrag an die Aachener Firma, die bis dahin noch recht wenig Bekanntheit erlangt

At the same time, the construction of the network structure was expanded, at times consisting of around 80 industrial partners. The automotive partner companies played a major role in the development of the StreetScooter. The success of the project would not have been possible without this integration strategy. StreetScooter then made a name for itself a year later, when it made its grand appearance before an international audience. The original model "A12" (the first functional primotype) was presented at the IAA 2011 in Frankfurt. When the Deutsche Post became aware of the StreetScooter at this first public appearance and awarded a development contract to the Aachen company – which until then had achieved very little notoriety – the StreetScooter really took off.

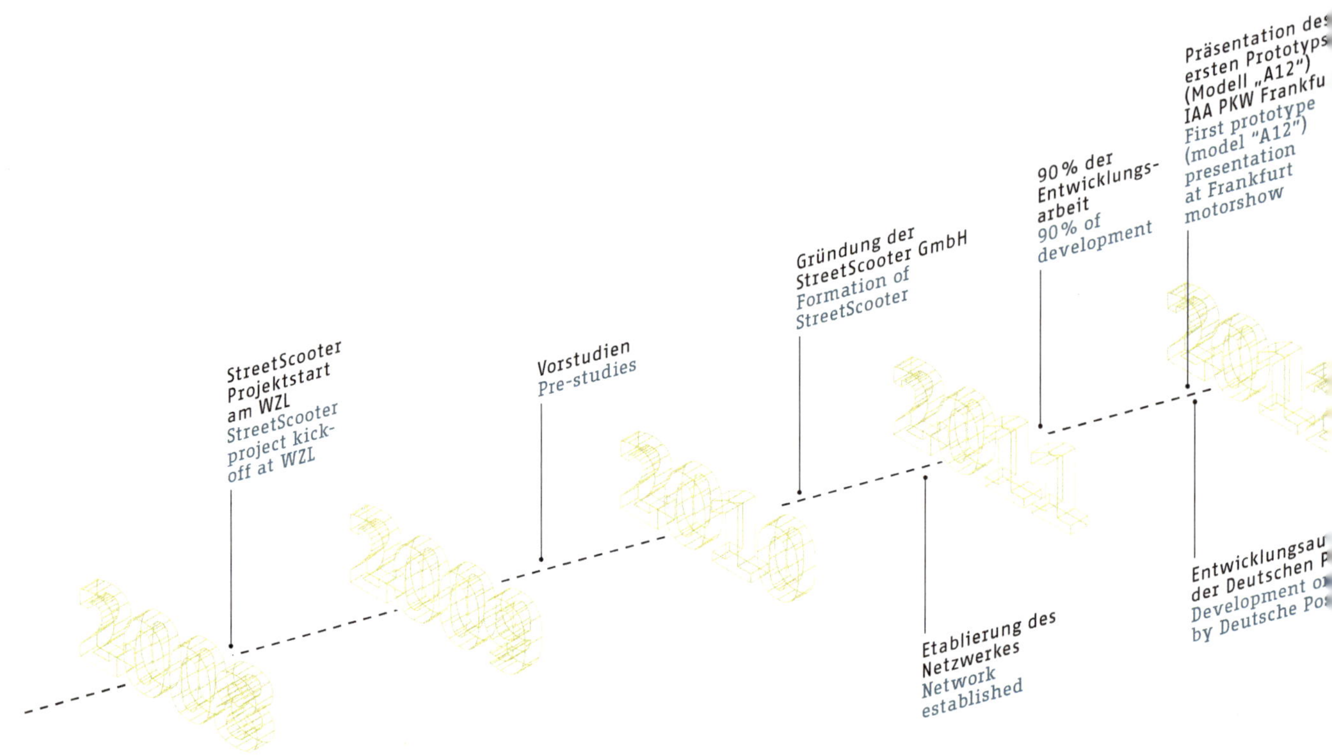

hatte, vergab, nahm der StreetScooter so richtig Tempo auf. Schon 2012 feierte der erste Post-StreetScooter namens „Work" auf der IAA-Messe für Nutzfahrzeuge in Hannover Premiere. Jetzt ging es Schlag auf Schlag: Zuerst wurde die Kleinserienproduktion der gelben „Nutzwerkzeuge" in Angriff genommen. Dann der Paukenschlag: Überzeugt von der großen Zukunft der kleinen Firma übernahm schon im Folgejahr 2014 die Deutsche Post die StreetScooter GmbH, um fortan unter der Regie des Aachener Start-up-Unternehmens eigene Fahrzeuge für die Brief- und Paketbeförderung zu produzieren. Nach dem Motto „außen gelb, innen grün" wurden nur ein Jahr später 200 gelbe, umweltfreundliche Fahrzeuge in Serie produziert. Ebenfalls 2015 eröffnete die StreetScooter GmbH

By 2012 the first Post-StreetScooter called "Work" premiered at the IAA for Commercial Vehicles in Hanover. After that, things happened in quick succession. First, the small series production of the yellow "utility tool" was tackled. Then the bombshell: Convinced of the great future of the small company, the Deutsche Post acquired StreetScooter GmbH the very next year (2014), to henceforth produce its own vehicles for mail and parcel transport under the direction of the Aachen start-up company. 2015: Just one year later, under the motto "yellow on the outside, green on the inside", 200 yellow, eco-friendly vehicles were produced in series. Also in 2015 StreetScooter GmbH opened a test track for delivery vehicles on the grounds of the German-Dutch Avantis business park. In this

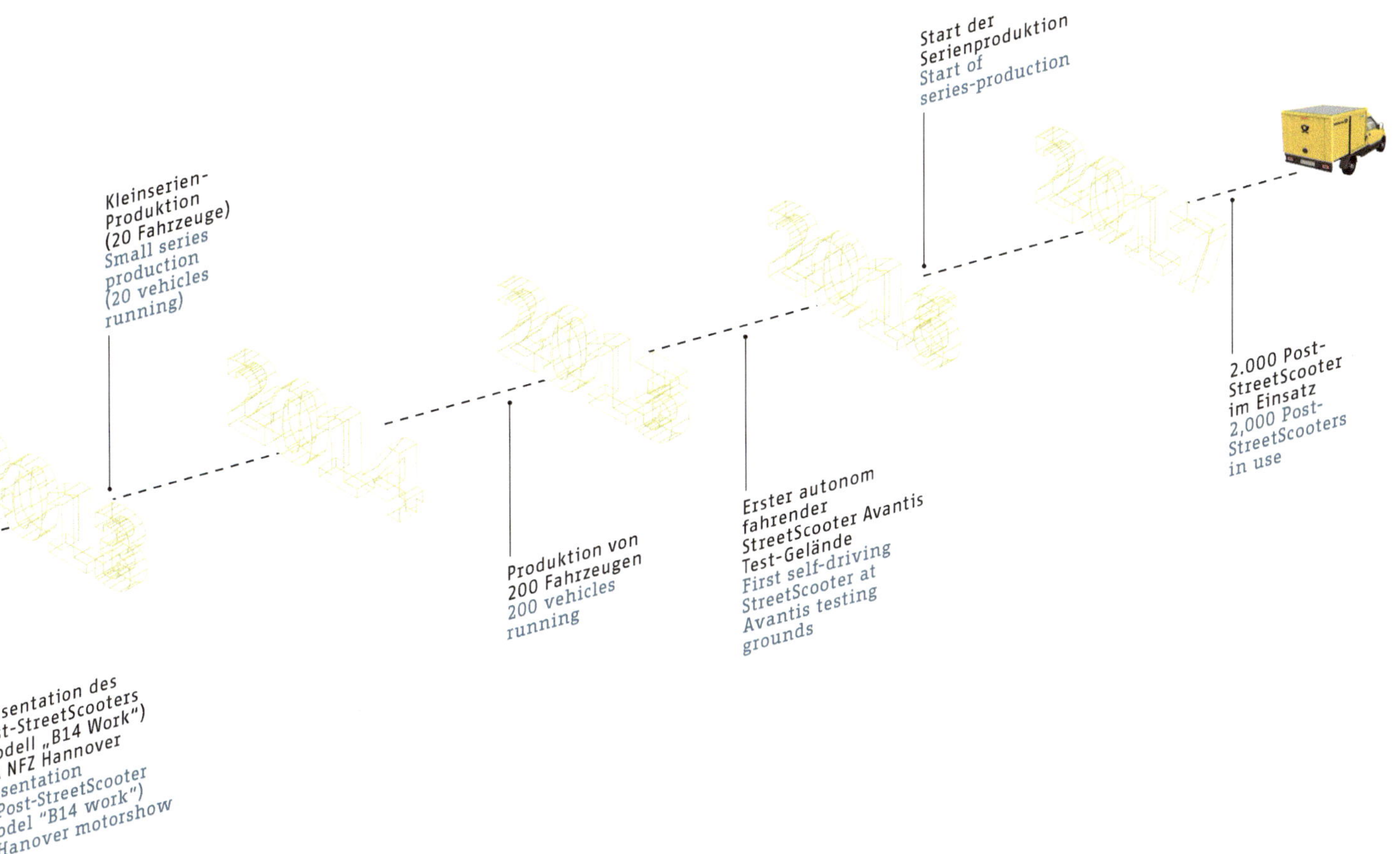

auf dem Gelände des deutsch-niederländischen Gewerbeparks Avantis eine Teststrecke für Zustellfahrzeuge. Auf der Fläche hatte das Bonner Mutter-Unternehmen zuvor den Grundstein für einen Innovationspark gelegt, auf dem seither zukunftsweisende Konzepte für die Zustellung getestet werden. Seit 2016 zeigen sich immer mehr des gelben StreetScooters aber auch im Straßenbild. Inzwischen (Stand Anfang 2017) sind rund 2.000 im täglichen Einsatz unterwegs und aus dem Kurzstreckenfahrzeug für die City wurde ein innovatives Zustellwerkzeug für den gewerblichen Einsatz, mit dem nun die Post ökologisch vorbildlich vorfährt.

area, the Bonn mother company had earlier laid the foundation for an Innovation Park, on which pioneering concepts for delivery are tested to this day. Since 2016, an increasing number of yellow SreetScooters have become an everyday sight in the streets. Right now (early 2017), approximately 2,000 are in daily use on the road, and out of the short-range vehicle for the city, an innovative delivery tool for commercial use has been born, with which the Deutsche Post has set a pioneering lead in ecological responsibility.

IMMER ALLES SCHON VORHER WISSEN, HEISST, WENIGER FREI ZU SEIN.

Prof. Achim Kampker
Geschäftsführer der Streetscooter GmbH
Executive Vice President E-Mobilität Deutsche Post DHL
CEO of Streetscooter GmbH
Executive Vice President E-Mobility Deutsche Post DHL

ALWAYS KNOWING EVERYTHING
ALREADY IN ADVANCE MEANS BEING
LESS FREE. THAT WOULD BORE ME.

MICH WÜRDE DAS LANGWEILEN.

4

DIE STREETSCOOTER-ENTWICKLUNGS-STORY

THE STREETSCOOTER DEVELOPMENT STORY

Nein, Design stand nie wirklich im Mittelpunkt der Entwicklung. Trotzdem ist der StreetScooter eine durchaus emotionale Kiste. Jedenfalls für die, die von Anfang an den Erfolg für möglich hielten.

No, the focus in the development was never on the design. Nevertheless, the StreetScooter is quite an emotional little jalopy. At least for those who believed it could succeed from the get-go.

FORM
FOLLOWS
FICTION

So schön kann elektrisch sein: der StreetScooter „Compact", wie er auf der IAA 2011 vorfuhr
Electric can be this beautiful: The StreetScooter "Compact" as was shown at the IAA 2011

Mit rationalen Argumenten war den Kritikern kaum beizukommen. Dass es möglich sein musste, ein Elektrofahrzeug mit einem extrem geringeren Investment in den Markt zu bringen, darüber hinaus in einer extrem kurzen Entwicklungszeit, daran glaubten die Kopfschüttler aus den Reihen der etablierten Marktteilnehmer eher nicht. „Das wird sowieso nichts", mussten sich die Initiatoren des StreetScooters lange anhören. Allein ihr Glaube an die Sache konnte den Zweifel besiegen und die Idee auf vier Räder stellen.

Während bei der klassischen Neuentwicklung eines Fahrzeugs rund sieben Jahre ins Land und die Investitionen dafür in die Milliarden gehen, schrumpfte bei der Entwicklung des StreetScooters das Zeitfenster auf kaum mehr als drei Jahre bis zum Serienlauf bei gleichzeitiger Minimierung der Kosten auf einen Bruchteil des herkömmlichen Investments. Was vielleicht wie Science-Fiction klingen mag, wird real nachvollziehbar, wenn man weiß, wie radikal die Produktentwickler der RWTH Aachen an den Prozess herangingen und ihn durchzogen – eine Wissenschaft für sich.

Winning critics over with rational arguments was never going to work. That it could be possible to bring an electric vehicle with an extremely low investment to market, and in an extremely short development time, was something that the naysayers from the ranks of the established market players had a hard time believing. "You're wasting your time" is something the initiators of the StreetScooter had to hear for a long time. But their faith in the project conquered any doubts and finally put the idea on four wheels.

While the traditional new development of a vehicle takes about seven years and billions in investments, with the development of the StreetScooter, the time window to series production shrank to little over three years, while minimizing the cost to a fraction of conventional investments. What might at first sound like science fiction, becomes perfectly understandable once you know how radically the product developers at RWTH Aachen approached and pulled off the process – a science unto itself.

LÖSUNGEN HEISSEN LÖSUNGEN, WEIL SIE NICHT AM STATUS QUO FESTHALTEN

SOLUTIONS ARE CALLED SOLUTIONS, BECAUSE THEY ACTS AS SOLVENTS TO THE STATUS QUO

Albert Einstein wurde einmal gefragt, wie er zu seinen Ideen käme. Er antwortete: „Ich taste mich heran."

So oder so ähnlich muss es auch in den Köpfen der StreetScooter-Vordenker zugegangen sein, als sie 2008 anfingen, über innovative Entwicklungsprozesse in der Automobilindustrie nachzudenken.

Das zugrunde liegende Problem, mit dem sie sich konkret zu beschäftigen begannen, beschäftigt die Automobilkonzerne und Zulieferer bereits seit Jahrzehnten: Über alle Klassen und Marken hinweg laufen den Herstellern die Entwicklungskosten und -zeiten davon.

Die zunehmende technologische Komplexität der Fahrzeuge, deren aufwendigere Entwicklung und Produktion ist das eine. Das andere ist das Mehr an Funktionen der Modelle, das die Marken neben dem Design benötigen, um sich zu differenzieren, getrieben vom wachsenden Wunsch nach Individualität der Autokäufer. All das drückt bei den Herstellern auf die Stimmung, denn es drückt auf ihre Bilanzen.

Zwischenfazit: Bedingt durch kurze Modellzyklen und zahlreiche Modellvarianten, die sich der nach Individualität strebende Konsument überall auf der Welt wünscht, die aber auch die Marke braucht, um im besten Fall weltweit einzigartig sein zu können, wird der Gewinn der Unternehmen in immer größeren Teilen von den Forschungs- und Entwicklungskosten gefressen.

Das Problem ist nicht neu und es hat einen Namen: Industrialisierungsprozess. Und der hat einen negativen Beigeschmack. Denn der Prozess dauert zu lange und kostet zu viel Geld. Time is money.

Albert Einstein was once asked how he came up with his ideas. He answered: "I just feel my way towards them."

That's more or less how we need to think when we reflect on innovative development processes in the automotive industry.

The underlying problem we need to specifically address has occupied automobile companies and suppliers for decades now: Manufacturers bear the costs and the time outlay for development across all classes and brands.

The increasing technological complexity of the vehicles, their elaborate design and production is one thing. The other is an increase in features of the models, which the brands need in addition to the design in order to differentiate themselves, driven by car buyers' growing desire for individuality. All of which depresses the manufacturers, because it depresses their profit.

Interim conclusion: Company profits are increasingly being eaten up by R&D costs for two reasons. On the one hand, there are short model cycles and numerous model variants, which are driven by the increasing desire on the part of consumers for individuality. And on the other, the consumers want the brand to be unique, and ideally to be so everywhere in the world.

The problem is not new and it has a name: Industrialization process. And it has a negative connotation. Because the process takes too long and costs too much money. Time is money.

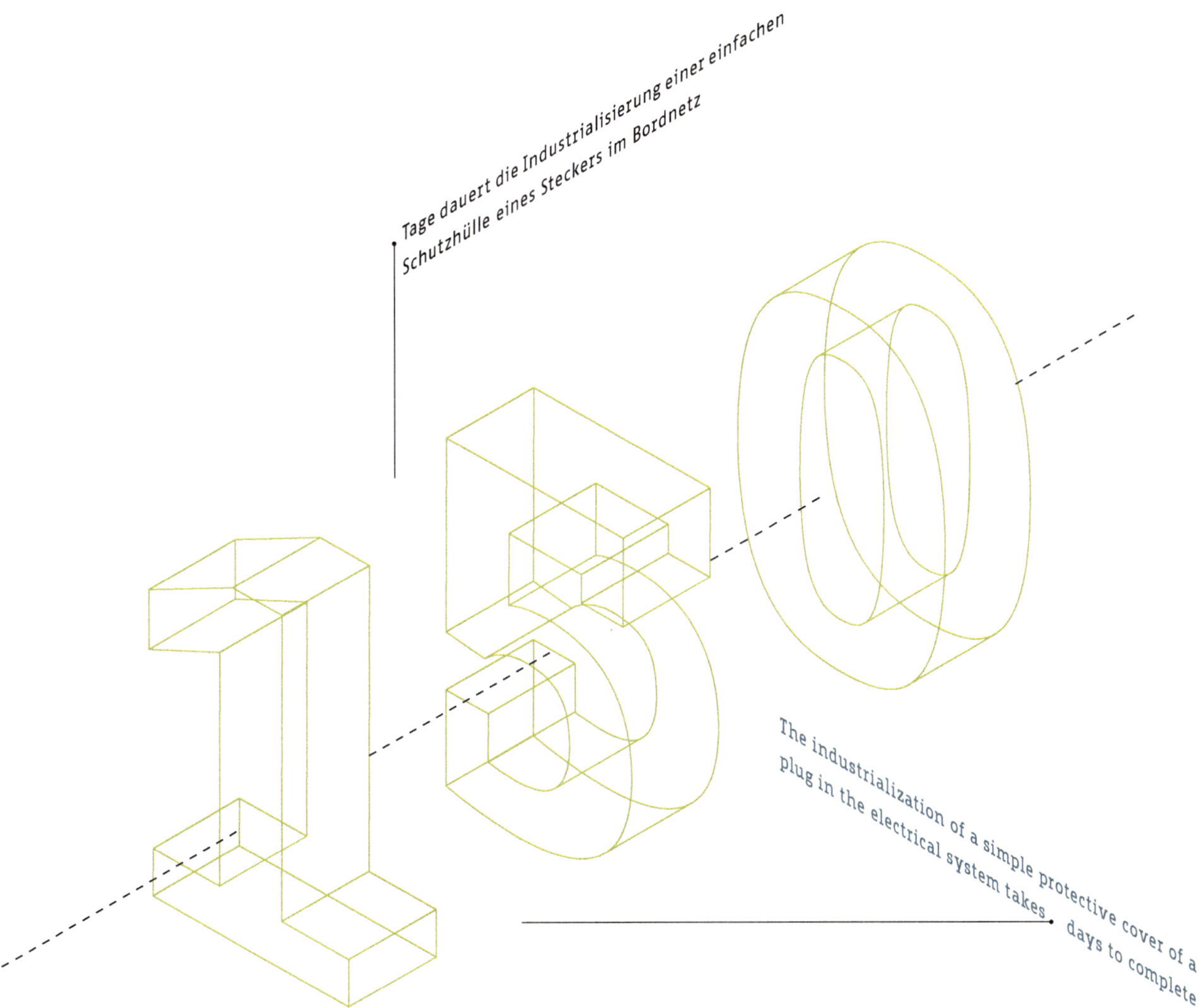

Tage dauert die Industrialisierung einer einfachen
Schutzhülle eines Steckers im Bordnetz

The industrialization of a simple protective cover of a
plug in the electrical system takes   days to complete

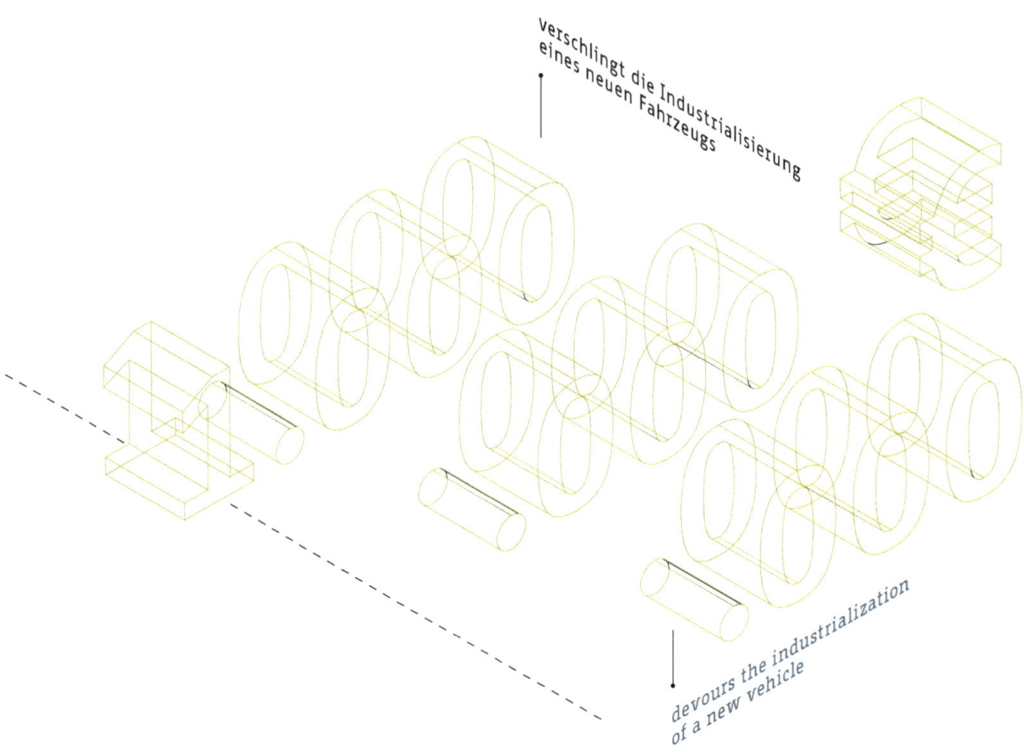

Perfektionismus bremst den Innovationsprozess

Zeit ist Geld. Diese Formel gilt für die Entwicklung eines Kleinteils, das irgendwo im Fahrzeug verbaut wird, bis hin zum kompletten neuen Fahrzeugmodell. Schon die Industrialisierung einer vermeintlich perferkten Schutzhülle eines Steckers im Kabelbaum dauert bis zu 150 Tage[9]. Das sind auf den ersten Blick 5 Monate. Überschaubar? Das ändert sich, wenn man bedenkt, dass das Jahr im Schnitt nur 250 Arbeitstage hat – also etwas mehr als netto 8 Monate – dann sind 5 Monate in Relation zu 8 bereits ein dreiviertel Jahr. Ein dreiviertel Jahr für einen einzigen Stecker! Vom Konzept bis zur Serienreife. Und von diesen Bauteilen gibt es tausende in einem Auto. Da ist es kaum verwunderlich, dass der Industrialisierungsprozess für ein komplettes, neu zu entwickelndes Fahrzeug Investitionen von 1 Mrd.[10] Euro verschlingt. Unterm Strich heißt das: Die Entwicklungskosten von der ersten Skizze bis zum Prototyp werden zum kritischen Erfolgsfaktor für jedes Unternehmen der Automobilindustrie.

Perfectionism slows down the innovation process

Time is money. This formula applies to the development of a small part that is installed somewhere in the vehicle, for instance, or all the way to a complete, new car model. Even the industrialization of a supposedly perfect protective cover for a plug takes up to 150 days[9]. So at first glance that comes to 5 months. Manageable? That changes when you consider that on average there are only 250 working days to a year – slightly more than 8 months net – since 5 months in relation to 8 is already nine months (3/4 of a year). Nine months for a single plug! From concept to production readiness. And a car has thousands of these components. So it is hardly surprising that the industrialization process for a complete, redeveloped vehicle can swallow up an investment of 1 billion[10] Euros. The bottom line is: The development costs from the first sketch to the prototype are becoming a critical success factor for any company in the automotive industry.

9 Quelle: Experteninterview
Source: Expert interview

10 Quelle: Best, Weth (2005) Geschäftsprozesse optimieren
Source: Best, Weth (2005) Geschäftsprozesse optimieren [Optimizing business processes]

Mit im Boot – oder besser gesagt: im Automobil – sitzen die Zulieferer. Schon vor etlichen Jahren hatten die Hersteller damit begonnen, die Entwicklungsaufgaben vermehrt den Zulieferern zu überlassen. Doch diese Lösung kann nicht als echte Lösung betrachtet werden. Denn das Problem wird nur verlagert. Zitat: „Schon jetzt schultern die Zulieferer laut der Unternehmensberatung Roland Berger gemessen am Umsatz höhere Entwicklungskosten als die Hersteller: Beim Hersteller sind es im Schnitt 5 % des Umsatzes, beim Zulieferer bis zu 8 %."[11]

Diese Zahlen sprachen schon 2002 für sich. Und auch heute hat das Problem immer noch Bestand und besteht auch nach wie vor auf Seiten der Automobilhersteller. Zitat: „Der globale Wettbewerb stellt Unternehmen vor neue Herausforderungen. Denn sie müssen in der Lage sein, schnell Produkte zu entwickeln und einzuführen, die den jeweiligen lokalen Marktbedürfnissen entsprechen. Einflussfaktoren wie regionale Markttrends, nationale Gesetzesauflagen und unterschiedliche Kundenwünsche führen dabei oft zum Konflikt zwischen kurzfristigen Zielvorgaben für das Management und langfristigen Innovationszyklen."[12] So sagt es Jochen Gleisberg, Partner von Roland Berger Strategy Consultants Ende 2013.

Entwicklungskosten auf der Überholspur

Die Fahrzeugentwicklung kann mit den Erwartungen des Marktes nicht Schritt halten. Die an der Entwicklung Beteiligten brauchen schlicht zu viel

And in the same boat – or car rather – are the suppliers. Several years back, the manufacturers had begun to increasingly leave development tasks to suppliers. But this solution can't be considered a true solution. Because all that does is pass the problem on. Quote: "According to the consulting firm Roland Berger, in terms of sales, suppliers shoulder higher development costs than manufacturers: For manufacturers, these are on average 5 % of their sales, whereas for suppliers it's as high as 8 %."[11]

Even as far back as 2002, these figures spoke for themselves. And the problem persists to this day, and still lies with the car manufacturers. Quote: "Global competition is posing new challenges for companies. Because they have to be able to quickly develop and launch products suited to local market needs. And when doing so, influencing factors such as regional market trends, national legal requirements and different customer requirements often lead to conflict between short-term targets for management and long-term innovation cycles."[12] Said Jochen Gleisberg, Partner at Roland Berger Strategy Consultants in late 2013.

Development costs on the fast track

Car development can't keep pace with the market's expectations. Those involved in development simply need too much time to efficiently reach the set goal. What's keeping them back, for example, are overly frequent coordination processes, overly

11 Quelle: Handelsblatt, 12.06.2002, Hohe Entwicklungskosten belasten/Automobilzulieferer begehren auf
Source: Handelsblatt, 6/12/2002, Hohe Entwicklungskosten belasten/Automobilzulieferer begehren auf [High development costs are causing a strain/Car suppliers are fighting back]

12 Quelle: Roland Berger Strategy Consultans, 04.09.2013, Vielfalt der globalen Märkte erfordert angepasste Produktentwicklung und -einführung
Source: Roland Berger Strategy Consultants, 09/04/2013, Vielfalt der globalen Märkte erfordern angepasste Produktentwicklung und -einführung [Diversity of global markets demands custom product development and launch]

Zeit, um das definierte Ziel, effizient erreichen zu können. Was sie bremst sind zum Beispiel zu häufige Abstimmungsprozesse, zu lange Werkzeugherstellung, zu viele involvierte Personen. Dabei werden sie von der Kostenentwicklung schlicht überholt.

Zurück zum Anfang und in die Köpfe der Vordenker: Die Idee des Return on Engineering hatte mit dem Glauben daran angefangen, dass sich mehr Kundenwert mit weniger Aufwand erzielen lassen muss – was zu beweisen war. Man begann mit der Entwicklung und Ausarbeitung der RoE-Idee, und zwar in einem überschaubaren Kreis von Netzwerkern. Denn eines war von Anfang an klar: Veränderungen sind mit Widerstand verbunden. Und dass dieses Motto umstritten sein würde, war ebenso von Beginn an klar: „Wir müssen früh scheitern, um schneller erfolgreich zu sein."

In diesem Satz steckte nicht nur ein Turbo für Innovationen; es wäre die Kraft des Neuen, die für ordentlich Bewegung in einem festgefahrenen Prozessdenken sorgen würde. Die Botschaft: Schneller aus ersten Schritten lernen, statt lange auf einen ersten Prototypen hinarbeiten.

long toolmaking and too many people involved. Here they are simply being overtaken by the cost of development.

Which brings us back to the start and to our thinking. Our idea of the Return on Engineering had begun with our belief that more customer value had been achieved with less effort. And we wanted to prove that this could be done. So we started with the development and elaboration of our RoE idea – within just a small circle of networkers. Because one thing was clear to us from the start: Changes were connected to resistance. And we knew from the get-go that our motto would be controversial: "We need to fail early to succeed more quickly."

This phrase wasn't just a boost for innovation; it would be the power of the new, which would ensure proper movement in a deadlocked process mindset. The message: To learn from first steps more quickly, instead of spending a long time working on a first prototype.

SCHEITERE SCHNELL, FRÜH UND KOSTENGÜNSTIG

FAIL QUICKLY, EARLY AND CHEAPLY

Hatte Einstein eigentlich einen Führerschein? Jedenfalls hatte er die Antwort auf die Frage, wie man schneller und effizienter ans Ziel kommt. Frei nach Albert E. haben wir so unsere eigene Devise entwickelt: Wir müssen uns vortasten, wir müssen immer wieder und wieder ausprobieren und wir müssen auch bereit sein, etwas, in das wir Zeit und Arbeit gesteckt haben, rasch wieder zu verwerfen; einzig, um mit einem geringeren finanziellen Einsatz schneller zu einem Ergebnis zu kommen, das für den Autokäufer höchsten Wert hat.

Mit dem Scheitern tun sich Menschen allgemein schwer. Max Levchin ist fünfmal gescheitert: „Das erste Unternehmen, das ich gegründet habe, ist mit einem großen Knall gescheitert. Das zweite Unternehmen ist etwas weniger schlimm gescheitert, das dritte Unternehmen ist auch anständig gescheitert, aber das war irgendwie okay. Ich habe mich rasch erholt, und das vierte Unternehmen überlebte bereits. Nummer fünf war dann PayPal."[13]

Was man daraus lernen kann? Scheitern gehört zum Erfolg dazu. Doch wie bei vielen Dingen ist auch hier die entsprechende Einstellung gefragt. Denn nur wer das Scheitern nicht als Untergang sieht, kann es als Chance begreifen und Nutzen daraus ziehen. Klingt einfach. Ist es auch, wenn man erst einmal akzeptiert hat, dass man dazu seine Einstellung ändern muss. Doch das ist schwer. Denn der Widerstand lauert überall, auch in uns selbst.

Did Einstein even have a driver's license? Who knows? But he did have the answer to this question: How can you get to your destination faster and more efficiently? And we've freely developed our own motto from Albert E.: We have to feel our way forward, we have to try things again and again and we must also be ready to quickly discard something in which we have invested time and effort. The goal: to more quickly achieve an outcome with the greatest value at a smaller financial investment.

People generally don't handle failure well. Max Levchin failed five times: "The very first company I started failed with a great bang. The second one failed a little bit less, but still failed. The third one, you know, proper failed, but it was kind of okay. Number four almost didn't fail. It still didn't really feel great, but it did okay. Number five was PayPal."[13]

What can we learn from this? Failure is part of success. But here, as with many things, you also need the right mindset. Only those who don't see failure as a downfall, can seize it as an opportunity and take advantage of it. Sounds simple. And it is too, as soon as you've accepted that you've got to change your mindset. But that's hard. Because the resistance to it lurks everywhere – even in ourselves.

13 Max Levchin, Mitbegründer und ehemaliger CTO (Chief Technical Officer) von PayPal
Max Levchin, Co-founder and former CTO (Chief Technical Officer) of PayPal

ES RECHNET SICH, MIT TEMPO DEM MISSERFOLG ENTGEGENZUSTEUERN

IT PAYS TO COUNTERACT FAILURE WITH SPEED

Wir Deutsche sind als Perfektionisten bekannt. Das hat uns in Wissenschaft und Technik weit gebracht. Nicht immer jedoch ist unser Fortschrittsdenken von Erfolg gekrönt. Das hat viel mit unserem Denken in alten Mustern zu tun. Wenn das Denken eine neue Form annehmen soll, tun wir uns alles andere als leicht. Etwa in Sachen effizienter Produktentwicklung.

Müssen wir uns also nicht selber fragen: Fehlt es uns an Mut, ja, an Radikalität?

Höhere Erträge durch schnelleren Markteintritt

Sucht man nach Pionieren, die radikal neue Wege gewagt haben, muss man über den großen Teich blicken: Amerika. Im Land der unbegrenzten Möglichkeiten ist offensichtlich unbegrenztes Denken leichter möglich.

Elon Musk, der Gründer von Tesla, riskiert mehr und erreicht mehr. Tesla hat einen Weg eingeschlagen, den deutsche Entwicklungschefs nicht gerne in Betracht ziehen: Musk

We Germans are known as perfectionists. This has taken us far in science and technology. And yet our concept of progress is not always crowned with success. This has a lot to do with our thinking in old patterns. When our thinking needs to assume a new form it's anything but easy for us to do. For instance, when it comes to more efficient product development.

So, don't we have to ask ourselves: Do we lack courage, are we afraid of being radical?

Higher income through faster market entry

If you're looking for pioneers who have radically ventured onto new paths, you'll have to look across the pond: America. In the land of boundless opportunity, boundless thinking is obviously easier.

Elon Musk, founder of Tesla, risks more and achieves more. Tesla has taken a path that German heads of development are loath to consider: Musk risked a quick market entry to secure his leading position. And in doing so, he simply factored

hat einen schnellen Markteintritt gewagt, um sich die Führungsposition zu sichern. Misserfolge hat er dabei gleich einkalkuliert. Er will den Markt für Elektroautos besetzen – jedenfalls so viel davon wie möglich. Und damit der Markt für seine E-Autos, an dem er partizipieren will, schnell größer wird, geht er sogar noch einen Schritt weiter: Er gibt seine Patente frei. Zukünftig darf die Konkurrenz seine Patente nutzen. Und das – so spekuliert Elon Musk – wird auch ihm nutzen.

Mister Tesla ging schon immer lieber vorneweg, auch wenn sein Weg nicht ohne Rückschläge war. Er hat den Tesla trotz seiner Kinderkrankheiten auf den Markt gebracht. Er hat Fehler, Misserfolge, Rückschläge in Kauf genommen. Seine visionäre Idee: Fehler schneller erfahren, um sie schneller eliminieren zu können. Tesla tickt eben anders. Und auch die Uhren gehen im Silicon Valley anders. Vor allem, was die Umsetzungsgeschwindigkeit betrifft.

Hier ist wohl auch der Grund zu sehen, warum Tesla im Markt für E-Autos mindestens eine Kühlergrillspitze voraus hat.

failures in. He wants to capture the market for electric cars – at least as much of it as possible. And to make sure that the market for his electric cars in which he wants to participate grows rapidly, he even goes one step further: He's sharing his patents for free. In the future, the competition will be allowed to use his patents. And that – Musk speculates – will also benefit him.

Mr. Tesla has always preferred leading the way, even if his way has been no stranger to setbacks. He brought the Tesla onto the market, despite its teething problems. He has accepted errors, failures and setbacks. His visionary idea? To identify errors more quickly, to better eliminate them. Tesla simply ticks differently. And the clocks in Silicon Valley tick differently. Especially when it comes to implementation times.

This is probably also where you can see the reason why Tesla is in the lead in the market for electric cars by at least one radiator grill.

BESSER TEUER VERKAUFEN ALS TEUER ERKAUFEN

BETTER TO SELL EXPENSIVE THAN BUY EXPENSIVE

Die Erkenntnis, dass die Industrialisierung eines Produktes sehr hohe zeitliche und damit finanzielle Aufwände verursacht, ist eine Tatsache, der man sich stellen muss. Doch wie kann der Aufwand heruntergeschraubt werden?

Martin Winterkorn, ehemaliger Vorstandsvorsitzender Volkswagen AG, machte 2014 klar: „Unsere Branche steht in den nächsten Jahren vor einem der größten Umbrüche seit Bestehen des Automobils. (...) Modellzyklen von sieben bis acht Jahren (müssen) deutlich kürzer werden."[14]

The realization that the industrialization of a product causes extremely high effort in time and by extension money is a fact we need to face. But how can the effort be scaled down?

Martin Winterkorn, former CEO of Volkswagen AG, made this clear in 2014: "In the next few years our industry will be facing one of the biggest upheavals since the creation of the automobile. (...) Model cycles of seven to eight years (must) be significantly shortened."[14]

14 Quelle: Manager Magazin (2014) Autoindustrie vor massivem Umbruch
Source: Manager Magazin (2014) Autoindustrie vor massivem Umbruch [Car Industry Facing Massive Upheaval]

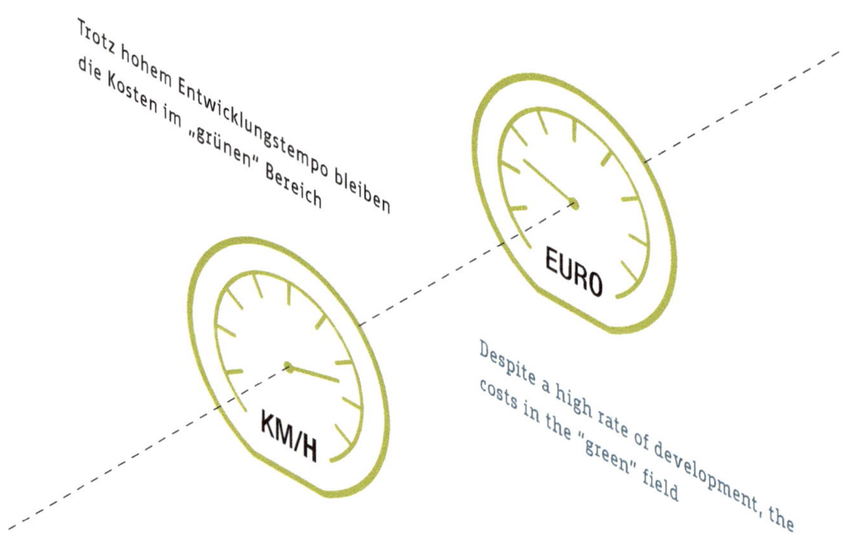

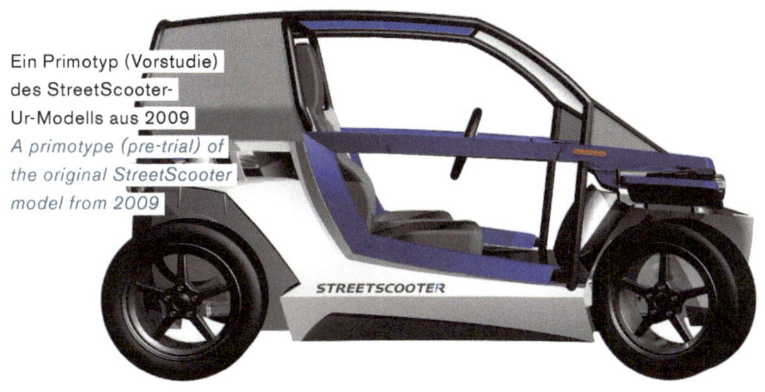

Ein Primotyp (Vorstudie) des StreetScooter-Ur-Modells aus 2009
A primotype (pre-trial) of the original StreetScooter model from 2009

Wer Umbrüche fordert, muss sein Denken umlenken. Das Ziel muss lauten: Prozessexzellenz. Langfristige Wettbewerbsvorteile sind nur auf diesem Weg erreichbar. Und zwar auf Dauer. Effiziente Industrialisierungsprozesse sind nämlich nicht nur ein einmaliger, sondern ein langfristiger Wettbewerbsvorteil. Der Grund: Effiziente Organisationsprozesse sind nicht leicht zu kopieren. Interne Prozesse sind von außen nur schwer zu analysieren. Sie sind selten ohne Anpassungen übertragbar. Ihre Implementierung kostet viel Zeit. Mit einem Satz: Prozess-Know-how sichert den langfristigen Wettbewerbsvorsprung[15].

Der Schlüssel für Effizienz im Industrialisierungsprozess

Steigende Funktionalität und Kundenindividualität werden in der Praxis durch immer höhere Industrialisierungsaufwände erkauft. Beispiel Volkswagen: Seit 2010 wurden 49[16] neue Modelle auf den Markt gebracht. Die Steigerungsrate betrug von 2010 bis 2014 rund 25 %. Im gleichen Zeitraum stiegen die Entwicklungskosten aber um 80 %. Die Margen pro Fahrzeug wurden deutlich kleiner.

Der Schlüssel zur Effizienzsteigerung im Industrialisierungsprozess ist also im Zusammenhang zwischen steigender Funktionalität (Kundenindividualität) und steigenden Industrialisierungsaufwänden zu suchen – und zu finden.

If you're calling for an upheaval, you've got to redirect your thinking. The goal must be: Process excellence. We postulate that long-term competitive advantages can only be achieved through process excellence. And in the long run at that. Efficient industrialization processes are not just a one-time, but a long-term competitive advantage. Why? Efficient organizational processes are not easy to copy. Internal processes are difficult to analyze from the outside. They are rarely transferable without modifications. Their implementation takes a lot of time. In a single sentence: Process expertise ensures a long-term edge on the competition[15].

The key to efficiency lies in the industrialization process

Increasing functionality and customer individuality in practice comes at an increasingly higher price in terms of industrialization. Take Volkswagen for example: Since 2010, 49[16] new models have been brought to the market. Between 2010 and 2014, this represented an increase of roughly 25 percent. But in the same period, development costs increased by 80 percent. The margins per vehicle shrank significantly.

The key to increasing efficiency in the industrialization process needs to be sought – and found – in the relationship between increasing functionality (customer individuality) and increasing industrialization efforts.

15 Vgl. hierzu Gaitanides (2007) Prozessorganisation
See also Galtanides (2007) Prozessorganisation [Process Organization]

16 Volkswagen Navigator 2010 und 2014 (ohne MAN, Ducati, Porsche)
Volkswagen Navigator 2010 and 2014 (excluding MAN, Ducat 1, Porsche)

KEEP IT SIMPLE: REDUCE IT TO THE MAX!

„Mache die Dinge so einfach wie möglich – aber nicht einfacher." Schon wieder Einstein. Der Mann war aber auch einfach genial, weil er genial einfach dachte. Aber man darf es sich auch heute nicht zu einfach machen. Zu einfach würde bedeuten, man macht so weiter wie bisher.

Status Schieflage: Industrialisierungsaufwand und Kundenwert

Man stelle sich eine Wippe vor: Industrialisierungsaufwand und Kundenwert stehen heute im Gleichgewicht. Das scheint auf den ersten Blick positiv. Der Haken: Mehr Kundenwert muss in der Praxis durch einen höheren Industrialisierungsaufwand erkauft werden. Was gebraucht wird, ist mehr Kundenwert bei weniger Aufwand. Die Kernherausforderung liegt darin, das Verhältnis zwischen Kundenwert und Industrialisierungsaufwand zu optimieren. Es geht darum, mit deutlich weniger Aufwand, also in kürzerer Zeit und mit weniger Investitionen, einen gleichen oder sogar höheren Kundenwert im Produkt zu erzielen. Der Kundenwert ist dabei als Ergebnis des Industrialisierungsprozesses zu sehen und der Industrialisierungsaufwand als Einsatz.

Das Verhältnis zwischen Ergebnis und Einsatz entspricht wiederum einem „Return on", analog einem Return on Investment oder Return on Capital Employed. Da die Tätigkeiten im Industrialisierungsprozess im Kern ingenieurstechnische Aufgaben sind, kann von einem Return on Engineering gesprochen werden. Der RoE kann hier aber auch als Lösungsansatz verstanden werden.

"Make things as simple as possible – but no simpler than that." Einstein again. But the man was simply brilliant, because his thinking was brilliantly simple. But by the same token, we mustn't allow ourselves to make things too simple either. Too simple would mean going on as we have before.

Status imbalance: Industrialization effort and customer value

Let's imagine a seesaw: Today industrialization effort and customer value are in balance. At first glance, that's a good thing. The catch: In practice, greater customer value takes a greater industrialization effort. What we need is more customer value with less effort. The core challenge is in optimizing the relationship between customer value and industrialization effort. The aim is to achieve an equal or even greater customer value in the product with much less effort, and by extension less time and with less investment. The customer value is to be seen as a result of the industrialization and the industrialization effort as a means to that end.

The relationship between results and effort in turn corresponds to a "return on", analogous to a return on investment or return on capital employed. Since the activities in the industrialization process are at root engineering tasks, we can speak of a Return on Engineering. But the RoE can also be understood as an approach to a solution. So let's get back to the image of the seesaw, or what we could also call a lever. Using the Return on Engineering, we are able to change the length of the lever in

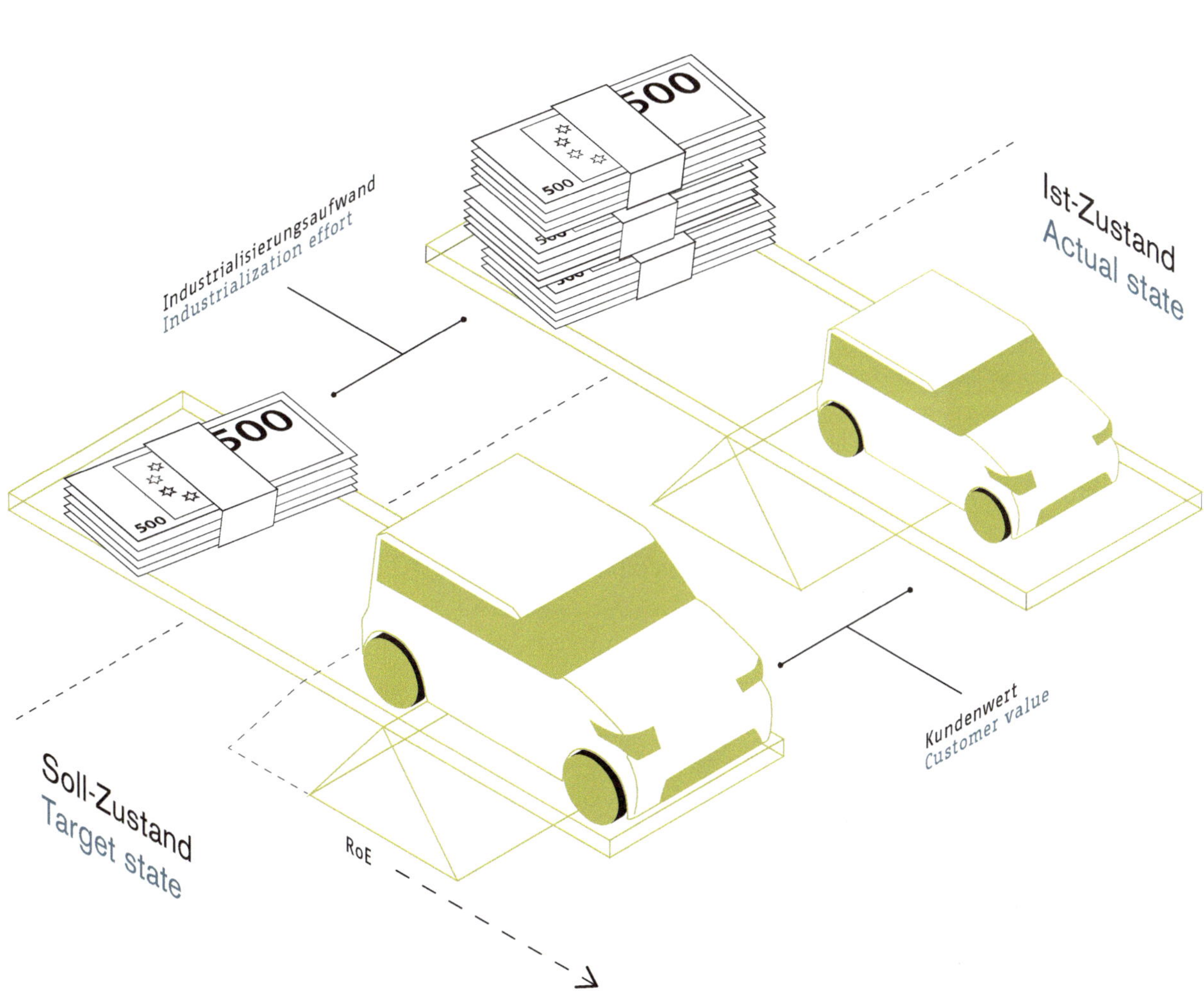
Industrialisierungsaufwand
Industrialization effort
500
500
500
Ist-Zustand
Actual state
Soll-Zustand
Target state
RoE
Kundenwert
Customer value

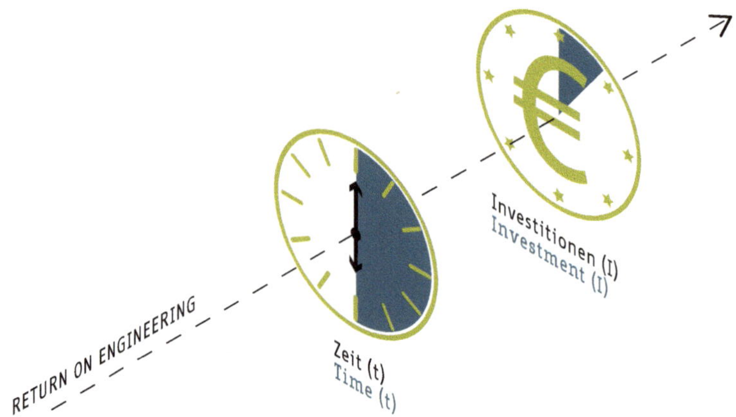

Man halte sich wieder die Wippe, die man auch Hebelbild nennen kann, vor Augen. Mithilfe des Return on Engineering ist man in der Lage, die Hebellänge zu verändern, um mit weniger Aufwand einen höheren Kundenwert zu erzielen. Die Vision: Der Industrialisierungsprozess ist in der Hälfte der Zeit (t) mit einem Zehntel der Investitionen (I) zu bewältigen.

Return on Engineering (RoE) = t/2 + I/10

Der RoE-Prozess braucht ein RoE-Management

Die Umsetzung von Veränderungsmaßnahmen im Unternehmen erfordert Investitionen und ein formales Management.

Wie bei jedem Veränderungsprozess ist zunächst ein Widerstand zu überwinden, um bestehende Strukturen zu verändern. So lässt sich auch der Hebel nicht einfach verändern, sondern es ensteht Reibung.

Am Anfang muss in den Veränderungsprozess investiert und die Umsetzung von einem formalen Management begleitet werden; vergleichbar etwa mit Lean Production. Wenn man RoE als Lösungsansatz versteht, um Strukturen im Industrialisierungsprozess zu verändern, dann ist auch hier analog zu Lean Production ein RoE-Management erforderlich – mit Trainings, Benchmarks und klaren Organisationsstrukturen.

order to achieve a higher customer value with less effort. Our vision: We need to be able to manage the industrialization process in half the time (t) with a tenth of the investment (I).

Return on Engineering (RoE) = t/2 + I/10

The RoE process needs an RoE management

The implementation of change measures in the company requires investment and a formal management.

As with any change process, there's a resistance to overcome at first in order to change existing structures. Meaning that the lever isn't easy to change: it's met with friction.

In the beginning there has to be investment in the change process and its implementation needs to be accompanied by a formal management; comparable to Lean Production. If we understand RoE as an approach for changing structures in the industrialization process, then just as in Lean Production, you need an RoE management – with training, benchmarks and clear organizational structures.

ERFOLGSTHESE 1

SUCCESS THESIS 1

Man spare sich den Prototypen

Status heute: Der Weg zum klassischen Prototyp ist weit, weil die Entwicklungszeiten lang sind. Steht der Prototyp endlich bereit, folgt erst die Erprobungsphase. Unterm Strich bedeutet das: hohe finanzielle Aufwände durch hohe Zeitinvestition in den Industrialisierungsprozess.

Status morgen: Zukünftig muss die Entwicklungsphase deutlich verkürzt werden. Oder anders gesagt: Sie muss stark segmentiert werden. Die klassische Entwicklungsphase wird gleichzeitig zur Erprobungsphase. Erfahrungen, die hier gesammelt werden, sind für den schnellen Lernprozess notwendig. Schnelligkeit versus Schwerfälligkeit. Statt sich auf den theoretischen Wissensaufbau zu stürzen, alles zu durchdenken, alles zu planen, ist es besser, in eine frühe Umsetzung zu gehen, um mit Primotypen Erfahrungen zu sammeln. Prototyp war gestern.

Fazit: Learnovational System

Klassische Prototypen sind nicht mehr notwendig, um ein entwickeltes Produkt zu testen. Stattdessen werden Primotypen für einen schnellen Lernprozess (Learning) früh im Innovationsprozess (Innovation) aufgebaut.

We save on the prototypes

Status today: The path to the classic prototype is long because the development times are long. Once the prototype is finally ready, then comes the testing phase. The bottom line is: high financial expenditure caused by high time investment in the industrialization process.

Status tomorrow: In the future, the development phase must be significantly shortened. In other words, they need to be highly segmented. The classic development phase simultaneously becomes the testing phase. The experiences we gather here are critical for a fast learning process. What drives us: Speed versus slowness. Rather than depending on theoretical knowledge building, thinking things through, planning everything, it's better to jump right into an early implementation in order to gain experience with primo types. Prototypes are a thing of the past.

Result: Learnovational System

Classical prototypes are no longer necessary for testing a developed product. Instead primo types are built early in the innovation process (Innovation) for a quick learning process (Learning).

ERFOLGSTHESE 2

SUCCESS THESIS 2

Wir verzichten auf fertige Lösungen

Kundenindividualität heute: Der Kunde kann heute aus einem breiten Produktprogramm sein nahezu individuelles Produkt konfigurieren. Dies bedeutet in der Praxis, dass immer mehr Produkte, Derivate und Ausstattungsoptionen angeboten werden. Das führt dazu, dass der Kunde schon bei einem Kleinwagen aufgrund der Vielzahl an Optionen zwischen 10^{12} Varianten wählen kann – Tendenz steigend. Gleichzeitig gibt es Produkte, die mit wenigen Varianten sehr erfolgreich sind. Beispiel iPhone: 2 Größen, 3 Farben und 3 Speichergrößen macht 18 Varianten. Und trotzdem hat jeder sein individuelles iPhone. Denn der Kunde kann sich eigene Anwendungen als Apps programmieren oder von Dritt-Anbietern zukaufen. Eine Art offene Schnittstelle ermöglicht die Individualisierung. Aber nicht nur in Software, auch in Hardware gibt es erfolgreiche Beispiele. Beispiel Nike: Hier kann der Kunde seinen Schuh individuell gestalten. Eine effiziente Prozesskette ermöglicht hier die kundenindividuelle Produktion.

Es geht also um effiziente Individualisierung eines Standardprodukts. Denn schon heute entwickeln bis zu 40 % der Kunden ihr Produkt weiter und kreieren eigene Derivate[17]. Ziel ist es, das Softwareprinzip auf Hardware zu übertragen, also neue Funktionen nachträglich „updaten" zu können, also ein „update-fähiges" Auto.

We don't use ready-made solutions

Customer individuality today: Today, customers can configure their practically individual product from a wide range of product options. In practice, this means that more and more products, derivatives and equipment options are offered. The result is that even in a subcompact car, customers can choose between 1012 variants thanks to the variety of options, and this is the growing trend.

At the same time, there are products that are very successful with very few variants. Take the iPhone: 2 sizes, 3 colors and 3 memory sizes gives you 18 variants. And yet everyone has their own individual iPhone. Because the customer can program their own applications as apps or buy from third-party providers. One way that open interface makes individualization possible. But it's not just in software that you've got successful examples: you've got them in hardware too. Take Nike: Here customers can personally customize their shoe. An efficient supply chain enables customized production.

It's all about efficient customization of a standard product. Because today up to 40 % of the customers continue to develop their products and create their own derivatives[17]. The aim is to transfer the software principle to hardware, i.e., to be able to subsequently "update" new functions, and in doing so create an "updatable" car.

17 Derivat, darunter versteht man eine Produktvariante, die von einem Basis-Produkt abgeleitet wird
Derivate: this refers to a product variant that is derived from a basic product

Fazit: Customer Engineering

Zur Vermeidung von Over-Engineering werden nicht alle Kundenanforderungen von vornherein im Produktprogramm abgebildet. Es geht vielmehr darum, ein Standardprodukt einfach und kostengünstig individualisieren zu können. Das Ziel: ein Auto, das sich nach Bedarf „updaten" lässt.

Result: Customer Engineering

To avoid over-engineering, not all customer requirements are mapped in the product range. It is more about being able to customize a standard product simply and inexpensively. The goal is a car that can be "updated" as needed.

ERFOLGSTHESE 3

SUCCESS THESIS 3

Man fokussiere auf den Konfigurationsprozess

Autos werden heute sehr stark auf ein Gesamtsystem ausgelegt und entwickelt. Das Fahrzeugkonzept wird definiert, die Anforderungen an die Teilkomponenten werden in der Zulieferpyramide top-down abgeleitet und die Komponentenentwicklung angestoßen. Schließlich werden die Teilkomponenten wieder zum Gesamtsystem integriert. Ein Auto wird so auf der ganzen Welt immer gleich verkauft. Der Kunde kann im Nachhinein kaum noch etwas ändern. Computer waren ursprünglich genauso aufgebaut. Dann tauchte der PC von IBM auf, der die Branche revolutionierte. Erstmals kam es zu einer Trennung zwischen Gesamtsystem und Modulen. Das Prinzip: Fertige Lösungen wurden zu einem Gesamtsystem konfiguriert. Standardschnittstellen ermöglichten die dezentrale Weiterentwicklung von Teilkomponenten. Oder anders ausgedrückt: Die Pyramide wird flach.

Standardlösungen aus dem Regal. Aber nicht von der Stange.
Es kommt darauf an, die Technologieentwicklung von der Produktentwicklung zu trennen, um in einem dezentralen Netzwerk auf Standardlösungen zurückgreifen zu können. Durch dezentrale Zuliefererstrukturen, deren Standardkomponenten wir nutzen, wird der Industrialisierungsprozess deutlich schneller und kostengünstiger. Der Fokus geht

We focus on the configuration process

Cars today are very much designed and developed based on an overall system. The vehicle concept is defined, the requirements for the sub-components are derived top-down in the supply pyramid and the component development is launched. Finally, the sub-components are reintegrated into the overall system. This way an identical car is always sold all over the world. There's hardly anything the customer can change afterwards. Computers were originally built the same way too. Then the PC from IBM came on the scene, which revolutionized the industry. For the first time there was a division between the overall system and the modules. The principle? Ready-made solutions were configured to an overall system. Standard interfaces made it possible to decentralize the development of sub-components. Or in other words: The pyramid becomes flat.

Standard solutions off the shelf. But not ready-made.
It is important to separate technology development from product development to be able to draw upon standard solutions in a decentralized network. Decentralized supplier structures, the standard components of which we use, makes the industrialization process much faster and cheaper. The focus then shifts away from the functional optimum of the overall system to the optimum at the

weg vom funktionalen Optimum des Gesamtsystems hin zum Optimum auf Komponentenebene. Die vergleichsweise geringe Differenz zum Funktionsoptimum wird kompensiert durch hohe Effizienzsteigerungen im Industrialisierungsprozess.

Fazit: Disruptive Network Approach

Die Produktgestaltung wird zu einem Konfigurationsprozess mit einem geringeren Fokus auf das Funktionsoptimum, der durch eine parallele Technologieentwicklung in einem dezentralen Netzwerk begleitet wird.

component level. The relatively small difference in the function optimum is offset by high efficiency in the Industrialization process.

Result: Disruptive Network Approach

The product design becomes a configuration process with a lesser focus on the function optimum, which is supported by a parallel technology development in a decentralized network.

PROTOTYP WAR GESTERN.
PRIMOTYP IST HEUTE.

PROTOTYPES ARE A THING OF THE PAST.
PRIMOTYPES ARE TODAY.

Besser schneller die erste Lösung entwickeln: Mit Blick auf den StreetScooter lautete das Ziel, möglichst schnell zu ersten Ergebnissen zu kommen statt ausgiebig und zeitraubend zu planen – diese Art von „Perfektionsstreben" dauert schlicht zu lange.

Rückblickend lässt sich heute sagen, dass das „Scheitern" ein wesentlicher Teil des Erfolgskonzeptes ist. Statt des hochgesteckten Zieles, auf den perfekten Prototypen zu warten, führt die Abkürzung zum Ziel. Wohl wissend, dass die erste Lösung noch Fehler haben wird und optimiert werden muss. Aber mit einer ersten 80%igen Lösung ist die Startbasis schon eine ganz andere und die Entwicklung kann deutlich schneller verlaufen.

So lassen sich früh Vorstufen zum klassischen Prototypen aufbauen und testen. Diese Vorstufen werden demnach Primotypen genannt. Auch im Projekt StreetScooter wurden bewusst früh erste Primotypen aufgebaut und zur Abfrage von Kundenanforderungen genutzt.

Better to develop the first solution more quickly. With a view to the StreetScooter, our goal was to get to the first results as quickly as possible, rather than extensively and time-consumingly plan – this kind of "perfectionism" just took us too long.

In retrospect, we can now say that for us the "failure" is an essential part of our success concept. Instead of the lofty goal, to wait for the perfect prototype, we prefer the shortcut to the goal; knowing that the first solution will still have errors and will need to be optimized. But with a first 80% solution, the launch base is already a very different one and the development can proceed much faster.

This way we can build and test precursors to classical prototypes early on. We call these precursors primotypes. We intentionally built early primotypes in the StreetScooter project too and used them for querying customer requirements.

Ein cooler Typ: Der „Compact" macht auch auf der Teststrecke eine blendende Figur

A cool guy: The "Compact" makes a dazzling impression on the test track

„DER STREETSCOOTER IST DER ROLLENDE BEWEIS DAFÜR, WIE MAN BEZAHLBARE ELEKTROMOBILITÄT AUF DIE STRASSE BRINGEN KANN."

"THE STREETSCOOTER IS THE ROLLING PROOF OF HOW AFFORDABLE ELECTROMOBILITY CAN BE TAKEN ON THE ROAD."

Fabian Schmitt
CTO, Geschäftsführung StreetScooter GmbH, Aachen
CTO, Management StreetScooter GmbH, Aachen

Sie sind technischer Geschäftsführer einer sehr jungen Firma. Sehen Sie sich selbst als Start-up?

Fabian Schmitt: Wir haben kein Bälle-Bad und wir haben keine Free Pizza und wir sitzen auch nicht im Silicon Valley. Wir machen aber auch ein paar Sachen anders als ein DAX 30, klar. Viele sogar. So gesehen sind wir mehr noch ein Start-up. Aber man erfindet sich auch nicht mehr so häufig neu, wie das am Anfang war, was dann einen Start-up irgendwo ausmacht. Wann was genau vom Start-up zu einer normalen Firma führt, weiß ich nicht, das verstetigt sich ganz automatisch.

Sie haben große Organisationsherausforderungen zu meistern, viel Zuwachs an Leuten, viel Dynamik.

Wir sind am Anfang nur drei Leute gewesen. Da musste man jede Funktion im Unternehmen irgendwie wahrnehmen. Das war eine echte Herausforderung. Man fängt mal irgendwie an, macht vieles, ändert wieder, passt an. Das ist nie perfekt. Dann entwickelst du weiter, neue Leute müssen integriert werden. Daneben musst du immer schauen, dass dieser Gründer-Spirit weiterlebt. Gerade am Anfang lebt so ein Unternehmen extrem von der Motivation der Menschen. Das ist schon sehr Start-up.

Hier trägt niemand Business-Look. Krawatte tragen auch relativ wenige.

Ob das jetzt ein Start-up ausmacht oder nicht? Kommt immer auf die Definition an. Wir sind unterwegs auf einem Weg. Aber es festigt sich auch. Man muss gewisse Sachen nicht täglich neu erfinden, wie das am Anfang gewesen ist.

Sie sind über die Start-up-Phase drüber weg, wenn man ehrlich ist. Schon erwachsen?

Ja. Aber da klingt so ein bisschen Negatives heraus. Als sei es besser, aus Kinderschuhen raus zu sein. Ist es im Prinzip ja auch. Wobei, man muss immer darauf achten, dass man das Positive aus dieser Phase mitnimmt und versucht, es am Leben zu halten.

Den Schwung.

Sicher, darum geht es.

You are the Technical Manager of a very young company. Do you consider yourself a startup?

Fabian Schmitt: We do not have a ball bath and we do not have free pizza and we are not located in Silicon Valley. And of course we do a few things differently than a DAX 30. Many things actually. From this perspective, we are still a startup. But we do not reinvent ourselves as often as we did in the beginning, which is what somehow defines a startup. When exactly a startup becomes a normal company, I do not know. It happens completely automatically.

You have to overcome great organizational challenges, a significant increase in people, many dynamics.

Initially there were only three of us. We somehow had to assume every role in the company. That was a real challenge. You start somewhere, do many things, make changes, adjust. It is never perfect. Then you advance yourself, new people have to be integrated. In addition, you have to always make sure that the founder spirit remains. At the very beginning, such a company is extremely dependent on people's motivation. This is really starting up.

No one here has a business look. Only few people wear a tie.

Whether or not this is indicative of a startup, I don't know. It always depends on the definition. We are on a journey. But it also creates a bond. You do not have to re-invent certain things every day, as it was in the beginning.

You have passed the startup stage, if you are to be frank. Already grown up?

Yes. But that sounds a bit negative. As if it were better to have come of age. And that is basically what it is. You have to make sure that you take the positives with you from this phase and try to keep it alive.

The momentum.

Of course, that's the idea.

Um was genau?

Ganz ehrlich, wenn wir hier nur Leute mit wenig Motivation hätten, könnten wir den Laden zumachen.

Hätten Sie ihn gar nicht aufgemacht?

Das ist schon so, dass unsere Leute die Begeisterung dafür haben, sich richtig reinzuwerfen.

Es gibt aber schon hierarchische Strukturen. CEO, CTO, CFO, leitende Mitarbeiter ...

Ja, und wir duzen uns trotzdem alle, um mal das Klischee zu bedienen. Das hat sich schnell ergeben. Wir pflegen einen direkten Umgang. Und es gibt ein ganz klares Wir-Gefühl.

In gewisser Weise also schon etabliert?

Gewisse Aspekte sehe ich da schon. Wir haben für alles mittlerweile Prozesse. Nichtsdestotrotz ist diese spezielle Mentalität, dass man ständig Sachen auch hinterfragen sollte, geblieben. Wie kann man es noch mal besser machen? Und auch mit noch mehr Herzblut? Da müssen dann auch Prozesse geprüft werden dürfen. So gesehen sind wir schon noch ein bisschen Start-up. Aber wenn man uns von außen betrachtet, sind wir ganz klar auf dem Weg zum normalen Unternehmen.

Welcher Mitarbeiter passt gut zum Unternehmen StreetScooter?

Die StreetScooter GmbH ist ein junges, unkonventionell geführtes Unternehmen mit kurzen Kommunikationswegen, schnellen Prozessen und flachen Hierarchien. Begeisterung wird bei uns großgeschrieben und ist einfach Teil unseres Arbeitsalltags. Jemand, der zu uns passt, muss diese Werte teilen. Das ist uns wichtig. Ansonsten muss er oder sie natürlich Elektrofahrzeuge entwickeln wollen. Wir arbeiten alle an einem Stück Zukunft. Gleichzeitig sitzen wir hier auf historischem Industriegelände der Firma Talbot. Hier werden seit Generationen Eisenbahnwaggons gebaut. Wer lieber in einen Glaspalast möchte mit gestyltem Büro, der sollte zu

What exactly?

Quite frankly, if we had only people with low motivation here, we could close shop.

Would you not have opened it in the first place?

It is true that our people have the enthusiasm to fully commit themselves.

However, there are already hierarchical structures. CEO, CTO, CFO, senior staff ...

Yes, and to confirm the stereotype: we are still informal in our communication. This happened quickly. We have maintained a direct approach. And there is a very clear feeling of belonging.

So, in some ways already established?

Indeed, I see certain aspects. By now we have processes for everything. Nevertheless, this particular mentality, that one should constantly question things, has also remained. How can you make it even better? And with even more dedication? Processes also have to be examined. In this sense we are still a bit of a startup. But when viewed from the outside, we are clearly on the way to becoming the normal company.

What kind of an employee fits in well at the StreetScooter company?

StreetScooter GmbH is a young, unconventionally managed company with short communication paths, fast processes and flat hierarchies. Enthusiasm is of utmost importance and is simply a part of our environment. Someone who fits in well with us has to share these values. This is important to us. In addition, he or she should have a desire to develop electric vehicles. We are all working on a piece of the future. At the same time, we are located on Talbot's historic industrial site. Railway cars have been built here for generations. If you prefer a glass palace with a stylish office, you should go to an insurance company. Look at our offices. Some things seem improvised. We are incomplete. This is visible in every corner. This is where future and history meet. This makes this

einer Versicherung gehen. Schauen Sie sich die Büros bei uns an. Bei uns scheint manches improvisiert. Wir sind unfertig. Das zeigt sich an allen Ecken und Enden. Hier kommen Zukunft und Geschichte zusammen. Das macht auch diesen Ort so besonders. Diese riesigen Produktionshallen, diese alten, schönen Backsteingebäude und darin wird nun fürs nächste Jahrtausend an einer neuen Mobilitätsidee entwickelt und gebaut.

Was ist Ihr Aufgabenfeld konkret?

Ich bin CTO, technischer Geschäftsführer. Ich verantworte alles, was technisch das Produkt betrifft.

Chief Technology Officer, also die oberste technische Leitungsperson.

Genau. Nehmen wir den StreetScooter, das Produkt: Body, Compartement, Türen, Fahrwerk, Motor, Powertrain, Antrieb, Batterie, Innenraum, Thermo-Management – das ist mein Verantwortungsbereich.

Es gibt mittlerweile auch ein Fahrrad, ein Pedelec.

Es gibt drei Modelle von uns: Die Transporter-Modelle „Work" und „Work L", das „E-Bike", das ist das Pedelec und es gibt das „E-Trike", ein Dreirad, ein Fahrrad, das mehr laden kann als das E-Bike.

Mit dem StreetScooter haben Sie das Elektrofahrzeug weiterentwickelt?

Nein! Einspruch. Wir haben explizit ein neues Elektrofahrzeug für den Kurzstreckenverkehr entwickelt, denn Autofahrer legen täglich meist ohnehin nur Strecken bis 50 Kilometer zurück. Dass daraus letztendlich dann ein Transportfahrzeug wurde, kam durch den Einstieg der Deutschen Post. Auch die Zusteller der Post legen täglich nur kürzere Strecken zurück.

Aber warum unbedingt eine Neuentwicklung?

Weil es bis dato noch kein Elektrofahrzeug speziell für dieses Segment gab. Es gab ja schon in den 1980ern und -90ern Bestrebungen, Elektromobile auf den Markt zu bringen, mit

place so special. These huge production halls, these old, beautiful brick buildings. This is where a new mobility idea is now being developed and built for a new millennium.

What is your specific responsibility?

I am CTO, Technical Manager. I am responsible for all the technical aspects of the product.

Chief Technology Officer, the top Technical Manager.

Exactly. Take the StreetScooter, the product: body, compartment, doors, chassis, engine, powertrain, drive, battery, interior, thermal management - this is my area of responsibility.

There is now also a bicycle, a pedelec.

There are three models from us: the transporter models „Work" and „Work L", and the „E-Bike", which is the pedelec and there is the "E-Trike", a tricycle, a bike that can carry more load than the E-Bike.

Have you further developed the electric vehicle with the StreetScooter?

No! Objection. We have explicitly developed a new electric vehicle for short-haul traffic, because drivers usually only cover routes of up to 50 kilometers per day. The resulting transport vehicle came about with the introduction of the German Post. Even the mailmen and women only cover shorter routes each day.

But why was a new development necessary?

Because there has not yet been an electric vehicle specifically for this segment. There were already attempts in the 1980s and 1990s to bring electric vehicles onto the market, with lead batteries, then with sodium sulfur batteries, then in the third wave with lithium-ion technology. Why have these vehicles not been able to establish themselves? The biggest obstacle was and is the functional limitation, the range. Only 100 kilometers! This had settled in the minds. Clearly, it is

Bleibatterien, dann mit Natriumschwefelbatterien, dann in der dritten Welle mit Lithium-Ionen-Technik. Warum haben sich diese Fahrzeuge nicht etabliert? Der größte Hemmschuh war und ist die funktionale Einschränkung, die Reichweite. Nur 100 Kilometer! Das hatte sich in den Köpfen festgesetzt. Klar, es ist viel emotionaler, einen Personenwagen für den Individualverkehr zu entwickeln und auf die Straße zu bringen, als ein Nutzfahrzeug. Aber wenn der Personenwagen zu wenig Reichweite erreicht und zudem in der Entwicklung viel zu teuer ist, sodass er nachher auch für den Endverbraucher viel zu viel kostet, dann muss man mit dem Denken mal neu anfangen, oder?

Das war eine Notwendigkeit, die Sie sehr früh für sich erkannt haben?

Dass jetzt, oh, Wunder, auch alle anderen sagen: Oh, ja, das ist ja viel zu aufwendig an den Privatkunden zu adressieren, das haben wir für uns rechtzeitig erkannt. Wir haben gesehen, dass gerade in kleineren, mittleren Flotten die effektiven Einsatzgebiete für ein Elektrofahrzeug sind. Das war ein Learning. Der Privatkunde ist nämlich viel zu differenziert in seinen Wünschen. Und er will auch ein paar hundert Kilometer fahren und dann sein E-Fahrzeug auch schnell wieder laden können. So weit ist das Elektrofahrzeug für Otto-Normal-Fahrer eben noch nicht. Aber es wird kommen. Der Flottenkunde – Logistikdienstleister, kommunale Betriebe und ähnliche Unternehmen – dagegen fährt eben nur 40 Kilometer am Tag oder ein paar mehr. Und er kommt auch abends meistens wieder in die gleiche Home-Base und kann über Nacht laden.

Stichwort: Batterie-Kapazität?

Batterie-Kapazität und -Kosten. Solange der Einzelne die 500 Kilometer nicht wirtschaftlich fahren und in zehn Minuten vollladen kann, solange bleibt das E-Mobil ein Flottenthema für den gewerblichen Kurzstreckeneinsatz.

much more emotional to develop a passenger car for private transportation and take it on the road than a commercial vehicle. But if the range of the passenger vehicle is limited and it is also much too expensive to develop, so that the final product costs way too much for the consumer, then you have to go back to the drawing board, right?

That was a necessity that you recognized very early on?

We have recognized this in time – and surprise: others are seeing it now, too – that it is too complicated to target the private customer. We have seen that electric vehicle can be most effectively used in small and medium fleets. This was a learning curve. The desires of the private customer are far too differentiated. And he also wants to drive a few hundred kilometers and then be able to quickly recharge his electric vehicle. So far, the electric vehicle is not yet for the average driver. But it will get there. The fleet customer – logistics service providers, municipal companies and similar companies – on the other hand, only drive around 40 kilometers a day. And he also returns in the evenings to the same home base where he can recharge overnight.

Keyword: battery capacity?

Battery capacity and cost. As long as the individual cannot drive 500 kilometers economically and can fully charge in ten minutes, the e-mobile will remain a fleet vehicle for commercial short-haul.

Looking at electromobility overall. What is the development capacity for private use?

There was a public hype at one point, then there was much talk and little action; both politically and on the part of the industry. It was a disillusionment because everyone joined this wave and promised that in three years we would

Elektromobilität insgesamt betrachtet. Wie steht es denn um die Entwicklungsfähigkeit für den Einsatz im Privaten?

Es gab ja mal einen öffentlichen Hype, dann ist viel geredet, wenig gemacht worden; politisch wie auch von Seiten der Industrie. Es war eine Ernüchterung, weil alle in diese Welle reingegangen sind und den Leuten versprochen haben, in drei Jahren fahren wir alle elektrisch. Vollkommener Nonsens. Und da saßen die Leute da und haben sich gefragt: Ja, wo sind denn die Produkte, die bezahlbar sind? Die kosten ja alle viel mehr. Die können ja viel weniger. Nein, will ich nicht! Und dann war dieses Thema eigentlich wieder tot. Keine OEM[18] hat einen richtigen Business-Case draus gemacht, weil am Horizont keine Stückzahlen zu sehen waren. Schlussendlich steht und fällt ja alles mit dem Preis, vornehmlich dem der Batterie. Und mit hohen Preisen kann ich im Markt keine Stückzahlen absetzen.

Das ist das klassische Denken, wenn ich als Hersteller nur auf die Batterie schiele statt auf den ganzen Industrialisierungsprozess?

Richtig. Die StreetScooter GmbH wurde aus der Idee heraus gegründet, Elektromobilität bereits ab kleinen Stückzahlen wirtschaftlich attraktiv gestalten zu wollen. Der StreetScooter ist das Realität gewordene Konzept eines kostengünstigen Elektrofahrzeuges, weil Ingenieure vieler verschiedener Fachrichtungen ihren Gehirnschmalz in den Entwicklungsprozess gesteckt haben

Und das hatte nichts mit der Batterie zu tun?

Wir haben das Fahrzeug ganzheitlich betrachtet. Was wir brauchten, war ein völlig anderes Entwicklungsprozessdenken als das herkömmlich bekannte.

Das Innovative ist damit nicht der StreetScooter.

Das Innovative ist der radikal neu gedachte Entwicklungsweg. Der StreetScooter ist der rollende Beweis dafür, wie man mit einem radikal anderen Denkansatz bezahlbare Elektromobilität auf die Straße bringen kann.

all be driving electric vehicles. Total nonsense. And the people sat there and asked: where are the affordable products? They all cost a lot more. They do so much less. No I do not want them! And then this topic actually died down again. No OEM[18] has been able to make a real business case out there, because no numbers could be seen on the horizon. In the end, everything depends on the price, especially the price of the battery. And I cannot deliver any large quantities to the market at high prices.

Is this the classical thinking, if I, as a manufacturer, only focus on the battery, instead of on the whole industrialization process?

Correct. The StreetScooter GmbH was founded from the idea of wanting to design electromobility at an attractive price from a small number of units. The StreetScooter has realized the concept of a low-cost electric vehicle because engineers from many different disciplines have put their brain power into the development process.

And that had nothing to do with the battery?

We looked at the vehicle holistically. What we needed was a completely different process of development than the one conventionally known.

The innovation is thus not the StreetScooter.

The innovation is the radically new development path. The StreetScooter is the rolling proof of how you can bring affordable electromobility to the road with a radically different mindset.

„

ANYONE WHO SAYS THAT YOU ALWAYS HAVE TO FOLLOW HIS WAY, SHOULD BE REMINDED: VERY OFTEN GOING CROSS COUNTRY GETS YOU THERE JUST AS WELL.

WER SAGT EIGENTLICH, DASS MAN IMMER **SEINEN WEG** GEHEN MUSS,

QUERFELDEIN

HAT AUCH SCHON MEHR ALS EINMAL ZUM ZIEL GEFÜHRT.

Prof. Achim Kampker
Geschäftsführer der Streetscooter GmbH
Executive Vice President E-Mobilität Deutsche Post DHL
CEO of Streetscooter GmbH
Executive Vice President E-Mobility Deutsche Post DHL

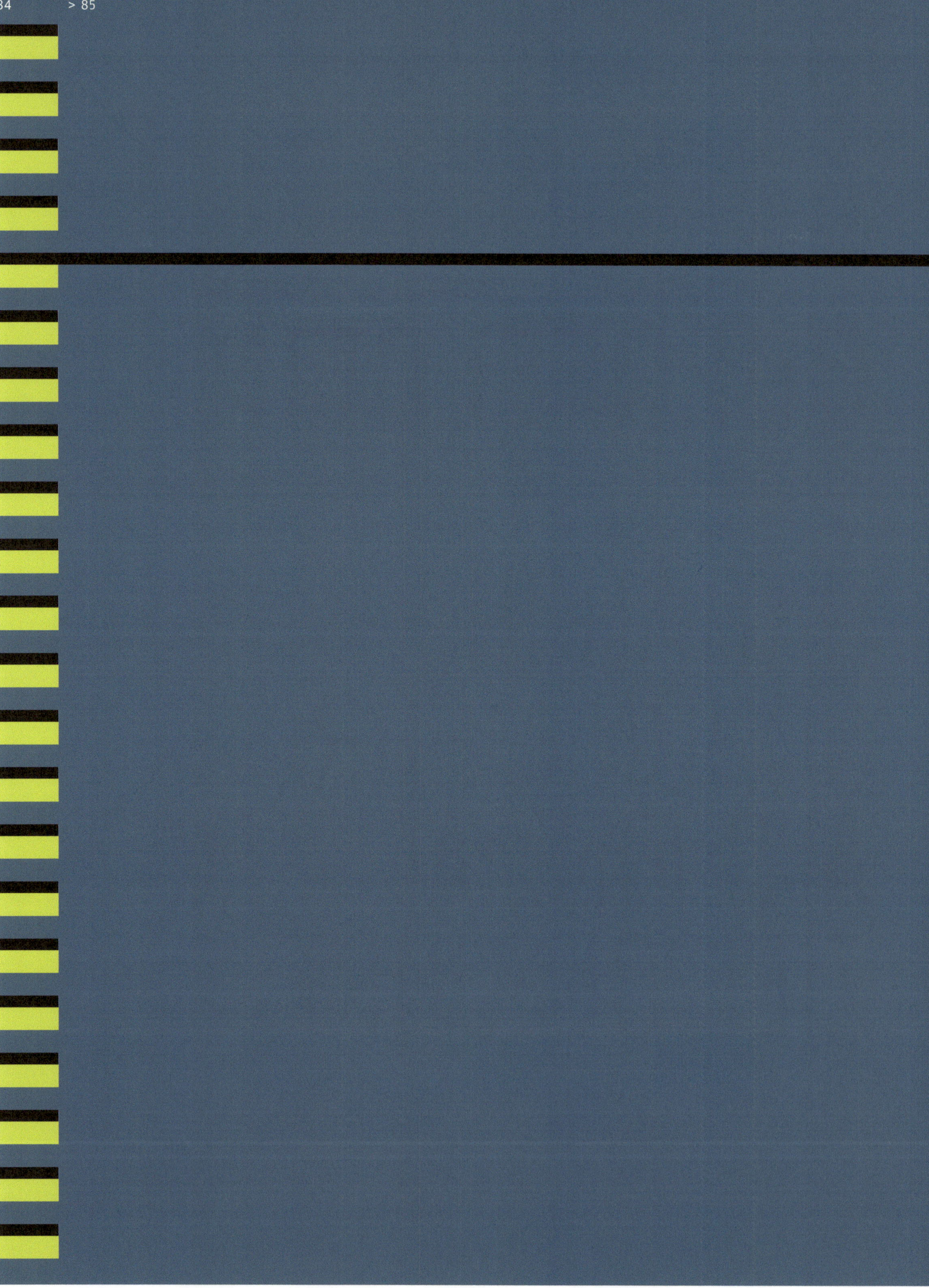

5

DIE STREETSCOOTER-PROTAGONISTEN-STORY

THE STREETSCOOTER PROTAGONISTS STORY

Keine Frage: Einer guten Idee ist es egal, wer sie hatte. Wenn aber aus seiner Idee eine Erfolgsgeschichte werden soll, dann ist es alles andere als gleichgültig, wer der Idee zur Seite steht. Die folgenden sechs Herren sind der personifizierte StreetScooter-Sixpack. Jeder von ihnen hat auf seine Weise Hirn, Herz und Muskeln für den StreetScooter spielen lassen.

It's obvious: A good idea is a good idea, no matter who thought of it. But for that idea to become a success story, it matters quite a lot who stands by it and supports it. The following six men are the StreetScooter six-pack personified. Each of them in their own way has flexed his brain, heart and muscles for the StreetScooter.

SECHS UNTERSCHIEDLICHE KÖPFE. EIN GEMEINSAMES ZIEL. SIX DIFFERENT HEADS. ONE COMMON OBJECTIVE.

Prof. Achim Kampker

Er ist Geschäftsführer der StreetScooter GmbH und Leiter des Geschäftsbereichs Elektromobilität der Deutschen Post AG. Entsprechend der Aufgabenbereiche, die er bekleidet, wirkt er eher vorstandsmäßig seriös. Trotzdem hat man bei ihm immer das Gefühl, er habe gerade die Ärmel hochgekrempelt und sucht nach der nächsten Aufgabe zum Anpacken. Wenn er dann auch noch zum Thema StreetScooter spricht, wird schnell klar: Er brennt für das Thema. Sein Motto: Geht nicht, gibt's nicht. Während der Entwicklung des StreetScooters ging es oft darum, spontan Lösungen für aufkommende Probleme zu finden. Kampker hatte in seinem Netzwerk immer einen Kontakt, der weiterhelfen konnte. So bei den Karosserien für die erste Vorserie. Als deutlich wurde, dass alle bestehenden Lieferanten nicht schnell genug liefern konnten, lief seine Telefonleitung heiß. Schließlich gab es einen Kontakt, der einen anderen Kontakt kannte, der schließlich einen kleinen Zulieferer kannte, der weiterhelfen konnte. Begeistern kann sich Prof. Kampker auch heute noch, wenn er an den Moment denkt, als sich die DHL in Aachen meldete. „Mit der Deutschen Post AG hatte StreetScooter zunächst einen verlässlichen Partner und schließlich einen wichtigen Investor gefunden, mit dem sich das Unternehmen substanziell weiterentwickeln konnte und der eine langfristige strategische Perspektive bietet. So konnte aus einer Idee etwas Reales entstehen – trotz vieler Bedenken und unzähliger Gründe, warum es eigentlich nicht gehen sollte."

He is the CEO of StreetScooter GmbH and Head of Electromobility Division at Deutsche Post AG. In keeping with the areas of responsibility in his remit, he comes off like a very serious member of the board. And yet, you always have this feeling with him as if he'd just rolled up his sleeves and was on the lookout to tackle the next challenge. So when he speaks about StreetScooter, it soon becomes clear: this is a topic near and dear to his heart. His motto: Nothing's impossible. During the development of the StreetScooter, it was often a matter of spontaneously finding solutions to emerging problems. Kampker always had a contact in his network who could help. For instance, with the bodies of the first pre-series. As it became clear that none of the existing suppliers were able to deliver fast enough, his phone line went white hot. In the end, there was a contact who knew another contact, who finally knew a small supplier who could help. Prof. Kampker can still get excited when he thinks back to the moment when DHL phoned up in Aachen. "With Deutsche Post AG, StreetScooter had initially found a reliable partner and ultimately an important investor, with which the company was able to significantly expand and which offered a long-term strategic perspective. This meant that from an idea something real could be created – despite many doubts and countless reasons as to why it really had no right to succeed."

Wer, wo, was innerhalb der StreetScooter-Story

▶ Übernahme StreetScooter als Hochschulprojekt von Prof. Schuh

▶ Akquisition der Shareholder zur Gründung der StreetScooter GmbH zusammen mit Prof. Schuh

▶ Aufbau des Unternehmensnetzwerks für die Entwicklung der ersten Prototypen

▶ Übernahme der Geschäftsführung seit 2010 und damit verantwortlich für den Unternehmensaufbau

▶ Akquisition der Deutschen Post als Kunde und Projektierung des Entwicklungsprojekts

▶ Entwicklung der zukünftigen StreetScooter-Strategie mit dem Start des Drittvertriebs

Who, where, what within the StreetScooter story

▶ Acquisition of StreetScooter as a university project by Prof. Schuh

▶ Acquisition of shareholders upon the founding of StreetScooter GmbH together with Prof. Schuh

▶ Building the company network for the development of the first prototypes

▶ Takeover of the management since 2010 and thereby responsible for the company structure

▶ Acquisition of Deutsche Post as a customer and project planning of the development project

▶ Development of the future StreetScooter strategy with the launch of third-party distribution

Jürgen Gerdes

Wer, wo, was innerhalb der StreetScooter-Story

► Längere, erfolglose Suche auf dem Fahrzeugmarkt nach kompaktem E-Zustellfahrzeug

► Begegnung mit Prof. Kampker

► Gemeinsame Idee, E-Kleinwagen und E-Zustellfahrzeug zusammenzuführen

► Begleitung und Förderung Konzept-Phase (u.a. Einbeziehung von 150 Zustellern zur Optimierung eines LabCars)

► Einsatz erster Pilotfahrzeuge am Firmensitz in Bonn, in der Folge Einsatz in immer mehr Schwerpunkt-Regionen (plus intensiver Öffentlichkeitsarbeit)

► Ernennung von Prof. Kampker zum Executive Vice President E-Mobility bei „Post – eCommerce – Parcel" der Deutschen Post AG

Who, where, what within the StreetScooter story

► Long, fruitless search in the automobile market for a compact E-delivery vehicle

► Meeting with Prof. Kampker

► Common idea, merging e-subcompact and e-delivery vehicle

► Concept phase guidance and support (e.g., involving 150 deliverers to optimize a LabCar)

► Use of the first pilot vehicles at company headquarters in Bonn, subsequently used in a growing number of key regions (plus intense public relations)

► Appointment of Prof. Kampker as Executive Vice President of E-Mobility at the "Post – eCommerce – Parcel" division of Deutsche Post AG

Jürgen Gerdes stieg schon 1984 nach dem Abitur bei der damaligen Deutschen Bundespost ein. Bis zur Berufung in den Vorstand der Deutschen Post AG war es freilich ein weiter Weg – 2007 kam er dort an. Ganz korrekt liest sich das heute so: Vorstand Deutsche Post, Ressort Post, eCommerce, Parcel. Damit ist er zuständig für das weltweite Brief-, Paket- und E-Commerce-Geschäft des Konzerns. Das alleine ist es aber nicht, was Jürgen Gerdes antreibt. Wenn er über seine Tätigkeit spricht, dann wird spürbar, dass er selbst auch ein Getriebener ist, aber ganz im positiven Sinne. Er will Dinge verändern, weiterentwickeln, neue Ziele erreichen. Für ihn liegt das Geheimnis von Können im Wollen. „Wenn wir Ziele setzen, dann reden wir über Wollen, und wenn wir Ziele erreichen, dann reden wir auch über Können." Und so war das auch beim Thema StreetScooter. Auf dem Markt gab es aus seiner Sicht kein passendes Elektromobil, das er in seiner Post-Flotte hätte einsetzen können. Also: selbermachen, ein eigenes, maßgeschneidertes Elektrofahrzeug entwickeln und bauen. Gewollt, getan. In Zusammenarbeit mit der Firma StreetScooter aus Aachen ging der Plan in Rekordzeit auf. Heute sind deutlich über 2.000 gelbe StreetScooter im Einsatz, zukünftig sollen es bis zu 30.000 sein. Aber auch abseits der Straße kann der StreetScooter gesichtet werden. Etwa im Stadion des 1. FC Köln. Das Maskottchen der Kölner, der Ziegenbock Hennes, wird bei Heimspielen des FC vom Kölner Zoo ins Fußballstadion mit einem leisen, emissionsfreien StreetScooter transportiert, den Jürgen Gerdes dem Club zur Verfügung gestellt hat. Und ehrlich: Da hat auch Hennes nichts zu meckern.

Jürgen Gerdes rose through the ranks at the former Deutsche Bundespost, having joined just after graduating high school in 1984, to his appointment to the Board of Deutsche Post AG in 2007 – he certainly came a long way! Quite justifiably, today, he is Chairman of Deutsche Post, Ressort Post, eCommerce, and Parcel. In that capacity, he is responsible for the Group's worldwide mail, parcel and e-commerce business. But that alone is not what drives Jürgen Gerdes. When he talks about his work, you can immediately sense that he himself is a driven man – but all in a good way. He wants to change things, develop them, and achieve new goals. For him, the secret to Can lies in the Wanting. "When we set goals, then we talk about Wanting and when we achieve our goals, we also talk about Can." And the same thing happened with StreetScooter. From his point of view, there was no suitable scooter on the market that he could have used in his fleet at Deutsche Post. Therefore, the answer had to be: do it yourself, develop and build your own, tailor-made electric vehicle. No sooner wanted than done. In cooperation with the company StreetScooter from Aachen, the plan worked out in record time. Today well over 2,000 yellow StreetScooters are in use, that number is planned to go up to 30,000 in the future. But the StreetScooter can also be spotted off the road. For instance in the stadium of Germany's Premier League Soccer's FC Köln. The mascot of the team from Cologne, Hennes the Goat, is transported on the team's home games from the Cologne Zoo to the soccer stadium in a quiet, emission-free StreetScooter, which Jürgen Gerdes has put at the team's disposal. And honestly: With a ride like that, Hennes has nothing to complain about.

Prof. Günther Schuh

Die Passionen, die man von Prof. Schuh kennt, wenn man ihn näher kennt, kann man alle auf einen Nenner bringen: Technik und Bewegung. Zu seinen Leidenschaften zählen Autos, Motorboote und das Fliegen. Da mag es wie die logische Fortsetzung einer Evolution klingen, wenn seine jüngste Passion nun auch ein Fortbewegungsmittel ist: der StreetScooter. Im selben Atemzug erklärt er, worauf der Erfolg eines Projekts basiert und dass nicht alleine kühle Technokratie entscheidend ist: „Jedes Projekt und letztendlich jedes Unternehmen lebt von einer guten Vision, die allen Beteiligten erklärt werden kann, die aber auch begeistert und inspiriert." Und über den Glücksfall Deutsche Post sagte er: „Mit dem Verkauf von StreetScooter an die Deutsche Post wurde der Weg in die Serienproduktion sichergestellt. Damit war gewissermaßen der ‚Proof of concept' aus der Hochschule vollbracht; der Beweis also, dass die erdachten Theorien und Ansätze auch in der Praxis umsetzbar sind und funktionieren." Prof. Schuh ist Inhaber des Lehrstuhls für Produktionssystematik am Werkzeugmaschinenlabor (WZL) der RWTH Aachen, Mitglied des Direktoriums am Aachener Fraunhofer-Institut für Produktionstechnologie IPT, Direktor des Forschungsinstituts für Rationalisierung (FIR e.V.) an der RWTH Aachen, Mitglied des Präsidiums der Deutschen Akademie der Technikwissenschaften acatec, Leiter RWTH Aachen Campus und Vorstandsvorsitzender der e.GO Mobile AG.

The passions, that Prof. Schuh is known for once you get to know him better, can all be boiled down to one common denominator: technology and movement. His passions include automobiles, motorboats, and flying. All of which may sound like the logical continuation of an evolution, if his recent passion is also a means of transportation: the Street-Scooter. In the same breath, he explains what the success of a project is based on and that it takes more than cool-headed technocracy. "Every project and ultimately every company thrives on a good vision, which can be explained to all those involved, but that also excites and inspires." And as to the Deutsche Post windfall, he said, "Selling StreetScooter to the Deutsche Post secured the path to series production. This was in a sense the 'proof of concept' for the university; the proof that the theories and approaches they conceived were also feasible in practice and that they worked." Schuh holds the Chair of Production Engineering at the Machine Tool Laboratory (WZL) at RWTH Aachen University, is member of the Executive Board at the Aachen Fraunhofer Institute for Production Technology IPT, director of the Research Institute for Operations Management (FIR eV) at RWTH Aachen University, member of the Presidium of the German Academy of Science ACATEC, head of the RWTH Aachen Campus and CEO of e.GO Mobile AG.

Wer, wo, was innerhalb der StreetScooter-Story

► Vordenker und Ideengeber
► Initiierung des StreetScooter-Projekts innerhalb der Hochschule
► Akquisition der Shareholder zur Gründung der StreetScooter GmbH zusammen mit Prof. Kampker
► Begleitung der Finanzierungsrunden und Unterstützung bei der strategischen Ausrichtung der StreetScooter GmbH
► Verhandlung mit der Deutschen Post AG zur Übernahme von StreetScooter

Who, where, what within the StreetScooter story

► Intellectual pioneer and initiator
► Initiation of the StreetScooter project within the university
► Acquisition of shareholders upon the founding of StreetScooter GmbH together with Prof. Kampker
► Supervision of the rounds of financing and assistance with the strategic direction of StreetScooter GmbH
► Negotiation with Deutsche Post AG on the acquisition of StreetScooter

Fabian Schmitt

Wer, wo, was innerhalb der StreetScooter-Story

▶ Der „Technologe" unter den Protagonisten

▶ Von Beginn an Fokussierung auf die technologische Entwicklung des Fahrzeugs

▶ Zunächst verantwortlich für technische Spezifikationen und technische Koordination des Unternehmensnetzwerks

▶ Dann verantwortlich für die LEG Gesamtfahrzeug und damit für Gesamtintegration der Einzelkomponenten

▶ Enge Betreuung des Aufbaus der frühen Prototypen und der Vorserienerprobung

Who, where, what within the StreetScooter story

▶ The "technologist" among the protagonists

▶ From the outset, focusing on the technological development of the vehicle

▶ Initially responsible for technical specifications and technical coordination of the company network

▶ Then responsible for the Lead Engineering Group (LEG) complete vehicle and thus for the total integration of the individual components

▶ Close supervision of the construction of the early prototypes and preproduction tests

Wenn man Fabian Schmitt trifft, denkt man: Ach ja, ein Gründertyp. Er kommt dem Klischee, dass alles in der Garage angefangen haben müsste, irgendwie am nächsten. Und tatsächlich. Rückblickend gibt es diese Garagenstory wirklich. Beim Aufbau des ersten StreetScooter-Prototyps hat sich Fabian Schmitt wochenlang in einer Garage auf dem Firmengelände eingesperrt und jede einzelne Schraube kontrolliert. Der Mann, der seiner Arbeit so besessen bis ins Detail nachgeht, ist heute CTO und Mitglied der Geschäftsführung der StreetScooter GmbH. Er zeichnet für alle Entwicklungsaktivitäten verantwortlich. Jeans und Kapuzenshirt trägt er heute noch lieber als Anzug und Krawatte. Das passt zu ihm. Seine Denke: Bestehendes in Frage stellen und neue Lösungen entwickeln. Nicht nur einen Standard abarbeiten und das Naheliegende realisieren. StreetScooter musste alles auf dem weißen Blatt Papier neu aufbauen und sich viele neue Kompetenzen erschließen. Und dabei wurde häufig aus der Not eine Tugend gemacht. Anstatt andere zu kopieren wurden neue Lösungen erdacht. Das sieht er auch im Kontext der Deutschen Post so: „Die Partnerschaft StreetScooter und Deutsche Post war aus Sicht der Entwicklung Chance und Herausforderung zugleich: Chance war ein klares Anforderungsprofil und Herausforderung das wahrscheinlich anspruchsvollste Anforderungsprofil überhaupt."

When you meet Fabian Schmitt, you think, Bingo: here's your typical start-up guy. And then somehow the cliché that everything probably started in his garage pops into your head. And it's true. In retrospect, there really is a garage story. When building the first StreetScooter prototype, Fabian Schmitt locked himself up in a garage on the company premises for weeks, checking every single screw. The man who pursues his work in such obsessive detail is now CTO and COO of StreetScooter GmbH. He is responsible for all development activities. To this day, he prefers wearing jeans and a hoodie to a suit and tie. That suits him. His way of thinking? Challenge what is and develop new solutions. Don't just work off of a standard and realize the obvious. StreetScooter had to rebuild everything starting with a blank sheet of paper and tap into many new skill sets. And in doing so often made a virtue of necessity. Instead of copying others, new solutions were thought up. He also sees this in the context of the Deutsche Post. "From a development perspective, the partnership between StreetScooter and Deutsche Post was at once both an opportunity and challenge. The opportunity was a clear requirement profile and the challenge was that these were probably the most demanding requirements possible."

Tobias Reil

Tobias Reil ist Produktionsleiter. In diesem Zusammenhang ist er verantwortlich für die Serienproduktion der Fahrzeuge und Pedelecs sowie für alle Prototypen und Vorserien. Reil ist jemand, der ganz vorne steht, jemand, der pragmatisch und unkompliziert Lösungen findet. So hat er sein Büro von Anfang an aus dem Bürogebäude heraus direkt in die Produktion verlegt. Ihm geht es nicht um das vermeintlich komfortable Office direkt bei der Geschäftsleitung, sondern er will direkt an der Produktionslinie sein. Gerade Tobias Reil hat mit der Produktion bewiesen, wie man in kurzer Zeit mit minimalen Investitionen eine professionelle Serienproduktion aufzubauen vermag. Mit der Post setzt er diesen Weg konsequent weiter um: „Was klein angefangen hat, ist inzwischen eine umfangreiche Partnerschaft. Längst wird nicht mehr nur das eine Post-Auto geliefert, sondern mit Pedelecs, Trikes und neuen Varianten des Fahrzeugs ein ganzes Produktprogramm produziert."

Tobias Reil is a production manager. In this regard, he is responsible for the series production of the cars and pedelecs as well as for all prototypes and pre-production series. Reil is someone who is very forthright, someone who finds solutions simply and pragmatically. Accordingly, he straightaway moved his office from the office building directly into the production area. He doesn't care about having a supposedly comfortable office directly next to management; he wants to be right at the production line. It was Tobias Reil himself who proved with the production how you could build a professional series production in short order and with minimal investment. With the Deutsche Post, he continues to consistently apply this approach. "What started out small, is now an extensive partnership. For some time now, we don't simply supply a Post-car, but with pedelecs, trikes and new versions of the vehicle we produce an entire product range."

Wer, wo, was innerhalb der StreetScooter-Story

- ▶ Zunächst für die Organisation des neuen Unternehmens und des Unternehmensnetzwerks verantwortlich, damit auch für die PLM-System-Einführung
- ▶ In frühen Phasen dann verantwortlich für die Entwicklung von Produktionskonzepten und Produktionsplanung
- ▶ Ab der ersten Vorserie dann verantwortlich für die Produktion
- ▶ Während sich Fabian Schmitt stärker auf die technischen Entwicklungsthemen fokussiert hat, hat sich Tobias Reil stärker auf die Strukturierung und Organisation des Unternehmens fokussiert

Who, where, what within the StreetScooter story

- ▶ First responsible for the organization of the new company and the company network, thus also for the PLM system implementation
- ▶ In early stages responsible for the development of production concepts and production planning
- ▶ After the first pre-production series, responsible for production
- ▶ While Fabian Schmitt focused more on the technical development issues, Tobias Reil focused more on the structuring and organization of the company

Dr. Christoph Deutskens

**Wer, wo, was innerhalb der
StreetScooter-Story**

► Vor der Gründung verantwortlich für die
Entwicklung von Produktionskonzepten

► In der StreetScooter GmbH zunächst verantwortlich für die LEG Body

► Später verantwortlich für die Koordination
der Forschungsprojekte

► Entwicklung der RoE Methodik durch Ana-
lyse des „StreetScooter Cases" und wissen-
schaftliche Arbeiten am Lehrstuhl PEM

**Who, where, what within the
StreetScooter story**

► Prior to the foundation, responsible for
the development of production concepts

► In the StreetScooter GmbH initially responsible for the LEG Body

► Later responsible for the coordination of
research projects

► Development of the RoE methodology by
analyzing the "StreetScooter Cases" and
scientific work at the PEM Department

In den Augen von Dr. Christoph Deuts-kens ist die gesamte StreetScooter-Er-folgsstory ein ewiges Lern- und Lehr-stück. „Welche Erkenntnisse können wir aus dieser Entwicklungsgeschichte zie-hen und wie lassen sich diese auf andere Unternehmen übertragen? StreetScoo-ter ist an vielen Stellen andere Wege gegangen. Das zu hinterfragen, zu ver-stehen und daraus Schlüsse zu ziehen, ist eine einmalige Chance." Deutskens ist Geschäftsführer PEM Aachen GmbH und PEM Consulting Mexico S.A. de C.V. PEM steht als Kürzel für: Production Engineering of E-Mobility Components. Klingt anspruchsvoll, ist es auch. Aller-dings muss man auch hier bereit sein, andere Wege zu gehen. Als es in der ers-ten Phase des StreetScooters um Pro-duktionsszenarien ging, mietete Deuts-kens mit einem Team eine alte, ziemlich runtergekommene Industriehalle an und baute dort über drei Monate eine komplette Produktion aus Pappe auf: Montagestationen, Fahrzeuge, Ladungs-träger etc. – alles war schließlich im Maßstab 1:1 aufgebaut und konnte zum Ausprobieren genutzt werden. Alles andere als von Pappe war der Einstieg der Deutschen Post. Als Deutskens und sein Team über den Produktionssze-narien brüteten, war das Thema noch nicht einmal in den kühnsten Träumen angedacht. Umso fantastischer wurde es dann: „Spannend an der Konstella-tion ‚StreetScooter-Deutsche Post' ist, wie aus dem ersten StreetScooter-Pro-totypen das Postauto wurde. Durch eine intensive Zusammenarbeit und Einbin-dung der Post-Mitarbeiter wurde effi-zient ein Fahrzeug im Purpose Design entwickelt."

In the eyes of Dr. Christoph Deutskens, the whole StreetScooter success story is an eternal lesson in learning and teaching. "What lessons can we draw from this development story and how can these be transferred to other com-panies? In many instances, StreetScoo-ter explored different avenues. To ques-tion this, to understand it, and to draw conclusions from it is a unique oppor-tunity." Deutskens is CEO of PEM Aachen GmbH and PEM Consulting Mexico SA de C.V. PEM is an abbreviation for Produc-tion Engineering of E-Mobility Compo-nents. Sounds ambitious, and so it is. However, you have to always be ready to try new approaches. When it came to production scenarios in the first phase of the StreetScooter, Deutskens and his team rented an old, rather run-down factory building, where they spent three months building a complete cardboard production. Assembly stations, vehicles, cargo carriers, etc. – everything was ultimately built on a 1:1 scale and could be used to try things out. Deutsche Post's arrival, however, was anything but lightweight. As Deutskens and his team were poring over their production scena-rios, this possibility would never have occurred to them even in their wildest dreams. That's why it was all the more fantastic when it actually happened. "The exciting thing about the 'Street-Scooter-Deutsche Post' constellation is how the first StreetScooter prototype became the Post-car. Through intensive cooperation and integration of the postal employees, a vehicle was developed efficiently in purpose design."

„ICH GLAUBE, DASS WIR STRUKTUREN AUFGEBROCHEN UND GEGEN GROSSE WIDERSTÄNDE NEULAND BETRETEN HABEN."

"I THINK THAT WE HAVE BROKEN DOWN BARRIERS AND HAVE ENTERED INTO UNCHARTERED TERRITORIES AGAINST TREMENDOUS ODDS."

Prof. Achim Kampker
Leiter des Lehrstuhls für Production Engineering
of E-Mobility Components (PEM) an der RWTH
Aachen und CEO StreetScooter GmbH und
Bereichsleiter Elektromobilität bei Deutsche Post
*Head of the Chair for Production Engineering of
E-Mobility Components (PEM) at RWTH Aachen,
and CEO of StreetScooter GmbH and Divisional
Manager of Electromobility at Deutsche Post*

Wo fängt für Sie der Moment an, in dem Neues gedacht wird?

Prof. Kampker: Ich glaube, dass es gar nicht so sehr darum geht, nur neue Ideen zu haben, sondern die Ideen, die es schon gibt, sinnvoll zu kombinieren und tatsächlich umzusetzen.

Was kann einer guten Idee gefährlich werden?

Es gibt so wahnsinnig viele Ideen und kluge Leute und Dinge, die schon durchdacht worden sind. Aber oft hat dann der Mut, die Überzeugung – ich weiß es nicht – gefehlt, die Sache anzupacken und zu tun. Der Flaschenhals heißt Umsetzung, durch diese Engstelle muss die Idee durch, sonst kommt im wahrsten Sinne nichts raus.

Wie sehen Sie sich? Sind Sie mit dem StreetScooter zum Erneuerer geworden?

Ich sehe mich nicht wie der große Mann mit der Einzelidee. Ich sehe eher etwas und füge das in einem Puzzle zusammen. So entsteht ein neues Bild, etwas Anderes, etwas Besseres. Ich entdecke Neues durch Beobachtung, wie man es in der Natur macht. Ich finde immer wieder faszinierend, wie viele Dinge man bei Tieren oder Pflanzen findet.

Bionik?

Ja, genau. Man findet Vorbilder in der Natur, Tiere, die genau diese Mechanismen, nach denen wir suchen, seit Jahrtausenden, Jahrmillionen benutzen. Da sage ich mir dann: Mensch, dieses Beobachten ist es doch. Was haben wir als Menschen wirklich erfunden? Haben wir es nicht eher gefunden oder herausgefunden, dass es das schon gibt.

Sie sind Professor und leiten einen Lehrstuhl. Kann es sein, dass Sie auch Unternehmergene in sich tragen?

Ich fühle mich weder in der einen noch in der anderen Welt komplett zuhause. Es ist für mich die Kombination, es gehört für mich zusammen.

At what moment do you think of something new?

Prof. Kampker: I believe that it is not so much a matter of having new ideas, but of meaningfully combining those ideas that already exist, and actually implementing them.

What can turn a good idea into a dangerous one?

There are so many ideas and clever people and things that have already been thought through. But oftentimes the courage or maybe the conviction to tackle the matter and get it done is lacking. The bottleneck is called implementation, the idea has to go through this bottleneck, otherwise literally nothing comes out.

How do you view yourself? Have you become an innovator with the StreetScooter?

I do not view myself as the great man with the one and only idea. Instead, I see something and put it together to form a puzzle. This creates a new image, something different, something better. I discover new things by observation, the way it is done in nature. I always find it fascinating how many things you can find in animals or plants.

Bionics?

Yes, exactly. You can find models in nature, animals, which have been using exactly those mechanisms for which we are looking, for millennia, millions of years. Then I say to myself: man, this observation is the answer. What have we really invented as human beings? Didn't we simply find it or found out that it already exists?

You are a professor who is the head of an academic chair. Could it be that you carry entrepreneurial genes?

I do not feel completely at home in one world or the other. For me, it is the combination, they belong together for me. Research is not an end in itself. Research just for the sake of research is not enough for me.

Forschung ist kein Selbstzweck. Einfach nur das Forschen um des Forschens willen ist mir zu wenig.

Man muss Dinge auch testen, ausprobieren, schauen, was sie taugen. Ich sehe für mich das Thema Neues generieren immer im Anwendungszusammenhang. Und dazu kommt: Ich bin auch kein Einzelkämpfer. Man kann gar nicht alles alleine stemmen. Man braucht Mitstreiter.

Dann macht es für Sie Sinn?

Für mich muss das klar sein und dann macht es mir richtig Spaß.

So war das für Sie auch bei der Elektromobilität?

Das E-Auto als solches, das ist keine neue Erfindung, gibt es schon. Und das höre ich ja auch allenthalben. Aber das, was wir jetzt gemeinsam geschaffen haben, dass die Autos tatsächlich gebaut werden und mittlerweile erfolgreich im Einsatz sind, das muss man sehen. Insofern ist mit dem StreetScooter schon was Neues entstanden. Aber es ist eben nicht das neue Auto, sondern die andere Herangehensweise an die Entwicklung und Realisierung. Wir sind durch den Flaschenhals und jetzt läuft es.

Wie bewegen Sie Andere, wenn Sie alleine nicht genug sind, wie Sie selber sagen?

Grundsätzlich bin ich hier in meinem Umfeld auf Menschen getroffen, die vom Grundsatz her offen sind für die Idee und ähnlich ticken wie ich. Diese dann zu begeistern, indem man glaubhaft machen kann, es ist tatsächlich ein Mittun und nicht für mich tun, ist sehr wichtig. Dann muss man den Mitspielern auch immer Freiräume geben. Auch das ist wichtig. Auch wenn sie es dann teilweise anders machen, als ich das machen würde oder mir vorstelle, trotzdem ist das super und gut und das ist, glaube ich, der Weg, der tatsächlich alle weiterbringt.

You have to test things, try them out, generate new ones in the context of the application. And moreover: I am not a lone wolf either. You cannot challenge everything by yourself. You need comrades-in-arms.

Then it becomes meaningful for you?

For me it must be clear and then it is really fun for me.

Was this also the case for you with the electromobility?

The e-car as such is not a new invention, it already exists. And I hear that everywhere. But what we have now created together, the fact that the cars are actually built and are now successfully in use, is something to behold. In this respect, the StreetScooter is a new invention. But instead of being the new car, the approach to the development and realization is different. We're through the bottle neck and now it's up and running.

How do you involve others when you alone are not enough, as you say?

Basically, here in my environment, I have encountered people who are in principle open to the idea and who tick as I do. To then inspire them, to be able to convince them to actually get involved and not just for my sake, is very important. You always have to give latitude to the other players. This is also important. Even if they do things partially differently than I would have done or imagined them, it is nevertheless great and good and I believe this is the approach that actually propels everything forward.

So: responsibility, identification, meaning.

The object is at the forefront, not me the person, for example. It is not about my reputation, about me, not about the fact that I have really pushed an idea this far.

Also Verantwortung, Identifikation, Sinn.

Die Sache steht im Vordergrund, nicht zum Beispiel meine Person. Es geht nicht um meine Reputation, um mich, nicht darum, dass ich ein Thema wirklich so weit vorantreibe.

Nochmal: Es geht mir um ein Thema und dafür möchte ich Mitstreiter gewinnen. Das gelingt mir gut bei all den Leuten, die noch nicht zu starr im Kopf sind. Dass einer sagt: „Es geht nicht!" kommt vor. Da aber jemanden zu überzeugen und abzuholen, das habe ich mir abgewöhnt, da verschwendet man von seinen eigenen Ressourcen viel zu viel, mit einem zu geringen Output.

Was sagen Ihnen dann Begriffe wie Hierarchie, Disziplin und Ordnung? Stellen sich da eher die Nackenhaare hoch?

Nein, ich glaube sogar, dass zur Umsetzung unseres Konzeptes eine gehörige Menge Disziplin gehört, auch Fleiß. Das sind schon Tugenden, die sehr sinnhaft sind. Ich frage mich eigentlich immer, wofür tust du das? Warum treibe ich eine Sache voran? Der entscheidende Unterschied ist zu wissen, wo man hin will und warum man etwas macht. Bei mir ist das Kernelement, dass ich etwas bewegen will, etwas, auf das ich stolz sein kann. Ich möchte mit dem, was ich tue, einen Keim pflanzen, aus dem etwas Besseres entsteht. Für alle. Aber nochmal: Fleiß, Disziplin und auch in gewisser Art und Weise Hierarchie gehören dazu. Und natürlich muss man führen. Es gibt da einen schönen Satz: Wer führen will, muss dienen.

Bei einigen Managern, die ich so erlebe, habe ich das Gefühl, die hätten auch gut in eine Zeit reingepasst, in der es noch Könige gegeben hat, die einfach nur Dinge tun, um Macht zu demonstrieren. Und das ist ja sowieso nicht der Boden für Neues, für Innovation, Weiterentwicklung und Kreativität.

Again: There is an idea and for this I would like to win fellow campaigners. I can do this well with all the people whose thought processes are not too rigid. There are occasions when some might say "it will not work!". I have, however, given up on the idea of persuading someone and taking them by the hand, because you waste too much of your own resources with too little output.

What do you think of terms such as hierarchy, discipline and order? Do they make the hair on your neck stand up?

No, I actually believe that the implementation of our concept requires a lot of discipline, also diligence. These are virtues that are very meaningful. I always ask myself why do I do this? Why do I push something? The key difference is knowing where you want to go and why you are doing something. For me, the core element is that I want to do something that I can be proud of. I want to plant a seed from which something better can develop. For everyone. But once again: diligence, discipline, and, to a certain extent, hierarchy are also a part of this. And of course you have to lead. There is a nice saying: He who wants to lead has to serve.

With some managers that I experience, I feel that they would have fitted in well at a time when there were still kings who just do things to demonstrate power. And that, in any case, is not the foundation for something new, for innovation, further development and creativity.

Is this a pioneering achievement that you have pulled off? Are you an evolutionary, have you advanced something or are you actually more of a revolutionary?

I believe we have broken down barriers and have entered unchartered territories against tremendous odds. And so this is somewhat of a pioneering achievement. It is not just an advancement but it was first something destructive in order to have become something constructive.

Ist das eine Pionierleistung, die Sie durchgezogen haben? Sind Sie Evoluzzer, haben Sie etwas weiterentwickelt oder sind Sie sogar mehr Revoluzzer?

Ich glaube schon, dass wir Strukturen aufgebrochen und gegen große Widerstände Neuland betreten haben. Und damit ist es so etwas wie eine Pionierleistung. Es ist nicht nur eine Weiterentwicklung, sondern es war zuerst etwas Destruktives, um zu was Konstruktivem zu kommen.

Was hat Sie denn bei diesem ganzen Prozess persönlich angetrieben?

Mich hat wirklich brennend interessiert, ob wir es schaffen. Es ging tatsächlich darum. Es stand eine These im Raum, verbunden mit einem Ziel. Ich habe eigentlich vom ersten Tag an daran geglaubt. Manchmal hatte ich auch Zweifel, aber in Summe habe ich immer daran geglaubt und gesagt: Das will ich jetzt nachweisen, dass das wirklich funktioniert. Und da war der Widerstand teilweise fast ein Antrieb für mich. Nach dem Motto: Jetzt erst recht.

Was genau wollten Sie beweisen? Es gibt ja diese RoE-Formel. Return-on-Engineering.

Der zentrale Punkt ist der, dass wir mit dem StreetScooter nach einer Umsetzungsmöglichkeit gesucht haben, um zu beweisen, dass das, was wir behaupten, tatsächlich funktioniert. So gesehen ist die Erfolgsgeschichte des StreetScooters „nur" ein Fallbeispiel. Mit einer Formel kann ich schön was behaupten und ein anderer kann das Gegenteil behaupten – und jetzt, was machen wir? Da aufzuhören war mir einfach zu wenig. Da wollte ich nicht stehen bleiben, sondern den Beweis antreten. Mein Motto war: Wir zeigen das jetzt erst mal an einem Beispiel, dass es funktioniert, und lernen gleichzeitig, wie es gehen kann. Denn was wir am Anfang nicht hatten, war die Blaupause, wie man es machen muss. Wir haben ja nur daran geglaubt, dass es Amerika gibt, wie ich dahin komme und wie weit das weg ist? Keine Ahnung.

What has driven you personally through this whole process?

I really wondered if we could make it. That was what it was really about. There was a thesis in space, connected with a goal. I actually believed in it from day one. Sometimes I had doubts, but in general, I always believed in it and said: I will now prove that this really works. And the partial resistance almost became a driving force for me. According to the motto: now more than ever.

What exactly did you want to prove? There is this RoE formula. Return on Engineering.

The main point is that we have been looking for a way to implement the StreetScooter, to prove that our idea actually works. In this regard, the success story of the StreetScooter is "just" a case study. With a formula, I can make a claim and another can claim the opposite – and now, what do we do? Giving up was not an option. That is not where I wanted to remain, I wanted the evidence to show. My motto was: we will now show for the first time, with an example, that it works and learn at the same time how it can work. Because what we did not have in the beginning was the blueprint of how to do it. We only believed that America existed, but how to get there, and how far away it was? No idea.

We also did not know what tools we required. We had the ideas and started to try them out.

You also did not know what to expect in the end?

Exactly, we did not know.

You could have drowned on the way.

The risk was not only latent, but it was always present, also with regard to financing. This pressure and the risk of failure were there, but I think this is always the case. The team moved from idea to idea without being certain if it would work. In this respect, this is probably the pioneer spirit. We were all aware of this. Also, that it could

Wir wussten auch nicht, was wir an Handwerkszeug brauchten. Wir hatten die Ideen und haben angefangen auszuprobieren.

Sie wussten auch nicht, was Sie am Ziel erwartet?
Genau, wussten wir nicht.

Sie hätten ja auch unterwegs absaufen können.
Das Risiko war nicht nur latent, sondern das war ständig da, auch was die Finanzierung angeht. Das heißt, dieser Druck und das Risiko des Scheiterns waren da, ohne das geht es, glaube ich, aber auch nicht. Das Team hangelt sich von Idee zu Idee und weiß nicht sicher, ob es funktionieren wird. Insofern ist das wahrscheinlich der Pioniergeist. Wir alle waren uns dessen bewusst. Auch dass es schiefgehen könnte und dass hinterher die Kritiker das Wort ergreifen: Das haben wir ja gewusst, dass das nicht funktioniert!

Können Sie einem Laien die Return-on-Engineering-Formel erklären?
Die Formel ist keine streng mathematische, sie soll aber Folgendes veranschaulichen: Es ist mit einem Bruchteil des herkömmlichen Investments möglich, ein Produkt auf den Markt zu bringen, und das auch noch in einer extrem kurzen Zeit.

Der Hintergrund ist der: Die Aufwände an Zeit und Invest, die für einen Industrialisierungsprozess getrieben werden, sind viel zu hoch. Unsere Frage lautete daher: Was muss ich tun, damit ich das signifikant anders machen kann? Die Methode wird durch die Formel illustriert. In ihr steckt der Leitgedanke zur Maximierung der Zeit- und Kosteneffizienz.

Signifikant heißt?
Zu einem Zehntel der Kosten in der Hälfte der Zeit.

Scheitern ist auch einkalkuliert?
Ja, aus Scheitern lernen. Ich glaube, es liegt in der Natur der Sache des Ausprobierens, dass

go wrong and afterwards the critics would be able to say: We knew that it would not work!

Can you explain the return on engineering formula to a layperson?
The formula is not strictly mathematical, but it is intended to illustrate the following: it is possible to get a product to the market, even in an extremely short amount of time and with a fraction of the conventional investment.

The background is the following: The expenditures in terms of time and investment for an industrialization process are far too high. So our question was: what do I have to do to make it significantly different? The method is illustrated by the formula. It is the guiding principle for maximizing time- and cost-efficiency.

Significant means?
At one-tenth of the cost, in half the time.

Is failure also taken into consideration?
Yes, learn from failure. I believe it is in the nature of the matter of trying out things that not every way can be the right one. You should perhaps take a few seconds to think about whether or not failure is the right term.

Experiment?
When I say I want to make it at a tenth of the cost in half the time, no one knows the solution. That is, I have to try it out and feel it out. The art is not to try one hundred ways, but to manage with trying ten. Failure is associated with it, but you cannot view it as such.

This is a playful, open approach to the subject, which of course isn't always affordable.
However, that is exactly what you have to be able to do, because it is overall significantly cheaper than the conventional way.

nicht jeder Weg der richtige sein kann. Da sollte man vielleicht nochmal ein paar Sekunden drüber nachdenken, ob Scheitern der richtige Begriff ist.

Experimentieren?

Wenn ich sage, ich will es für ein Zehntel des Aufwandes in der Hälfte der Zeit machen, kennt keiner die Lösung. Das heißt, ich muss ausprobieren und mich rantasten. Die Kunst liegt darin, nicht hundert Wege zu gehen, sondern mit zehn auszukommen. Scheitern ist damit verbunden, aber man darf es nicht so begreifen.

Das ist eine spielerische, offene Herangehensweise an das Thema, die kann man sich natürlich nicht immer leisten.

Doch, genau das muss man sich leisten, weil in Summe ist es deutlich günstiger als der herkömmliche Weg.

Mal weitergedacht: Kann man dieses Effizienzmodell auf andere Industrien übertragen?

Ja, definitiv. Man kann natürlich nicht alle Einzelheiten oder Details einfach kopieren. Ich glaube aber, die Grundmechanismen, wie zum Beispiel schnell in ein Thema reinzugehen, daran zu lernen, die kann ich übertragen.

Once again: Can this efficiency model be transferred to other industries?

Yes, definitely. Of course you cannot simply copy all of the particulars or details. I believe, however, that I can transfer the basic mechanisms, such as to quickly work my way into a topic and to learn from it. Those are transferable.

DIE LÖSUNG:
RETURN ON ENGINEERING

THE SOLUTION:
RETURN ON ENGINEERING

Prof. Achim Kampker
Geschäftsführer der Streetscooter GmbH
Executive Vice President E-Mobilität Deutsche Post DHL
CEO of Streetscooter GmbH
Executive Vice President E-Mobility Deutsche Post DHL

LACHSE SCHWIMMEN TAUSENDE KILOMETER

GEGEN DEN STROM. TROTZDEM HABEN SIE GEGENÜBER DEM MENSCHEN

6

DIE DREI STREETSCOOTER-LÖSUNGSBAUSTEINE

THE THREE STREETSCOOTER BUILDING BLOCKS

Auf den ersten Blick ist der StreetScooter ein E-Mobil: vier Räder, ein Chassis, ein elektrisches Antriebsaggregat. Doch das ist nicht die Innovation, die elektrisiert. Der StreetScooter ist vor allem ein erfolgreiches Beispiel für ein radikal anderes Denken in puncto Industrialisierung und Umsetzung von Entwicklungsprozessen. Dieser Erfolg basiert auf einer bestimmten Denkweise, einer konkreten Strategie sowie klar definierten Lösungsbausteinen und -prinzipien.

At first glance, the StreetScooter is a mobility scooter: four wheels, a chassis and an electric drive unit. But that's not the electrifying innovation. The StreetScooter is above all else a successful example of a radically different way of thinking in terms of industrialization and implementation of development processes. This success is based on a certain way of thinking, a concrete strategy, and clearly defined building blocks and principles.

EIN BERECHENBARER ERFOLG AUS 3 MAL 3 LÖSUNGSBAUSTEINEN = 9 LÖSUNGSPRINZIPIEN

A PREDICTABLE RESULT FROM 3 TIMES 3 BUILDING BLOCKS = 9 PRINCIPLES

Als der erste StreetScooter in Aachen aus dem „Versuchslabor" rollte, war etwas ganz Entscheidendes in Bewegung geraten: Aus dem Versuch, bezahlbare Funktionalität zu erschaffen, war ein realer Beweis für umsetzbare Elektromobilität entstanden. Es war der Anfang der StreetScooter-Erfolgsstory. Einmal in der Welt, nahm dieser Primotyp – damals noch ein kleiner Zweisitzer – immer mehr Fahrt auf. Heute ist aus dem exemplarischen Ur-StreetScooter ein im harten Post-Einsatz bewährtes und umweltfreundliches Zustellungsfahrzeug mit E-Antrieb geworden. Dabei ist nicht nur eine einmalige E-Mobile-Entwicklung vonstattengegangen, sondern es wurden im Laufe des Projekts auch 3 Lösungsbausteine, bestehend aus 9 Lösungsprinzipien, entwickelt. Lösungsbausteine und -prinzipien bilden im Zusammenspiel die methodische Basis für den Erfolg und stehen für die Umsetzung des Return-on-Engineering[19] (kurz RoE).

Auf den folgenden Seiten werden die 3 Lösungsbausteine und 9 Lösungsprinzipien der RoE-Methodik genauer beschrieben. Dabei wurde Wert darauf gelegt, die Inhalte so allgemein verständlich wie möglich zu formulieren, sodass sie auch auf andere Anwendungsfälle und Industrialisierungsprozesse übertragbar werden.

When the first StreetScooter rolled out of the "laboratory" in Aachen, something decisive had been set in motion: out of the attempt to create affordable functionality, a real proof of viable electromobility had arisen. It was the beginning of the StreetScooter success story. Once out in the world, this little primotype – then still a small two-seater – really started to pick up speed. Today, the proof of concept proto StreetScooter has become an environmentally friendly delivery vehicle with electric drive that has proven itself in demanding postal delivery practice. This was not simply a one-off e-mobile development, but 3 building blocks consisting of 9 principles were also developed during the project. Building blocks and principles, when in interaction, form the methodological basis for success and are responsible for the implementation of the Return on Engineering[19] (RoE).

On the following pages, the 3 building blocks and 9 principles of the RoE methodology are described in detail. Here emphasis was placed to formulate the content as generally understandable as possible so it can be transferred to other applications and industrialization processes.

19 Return-on-Engineering, von der RWTH Aachen am Lehrstuhl für Production Engineering of E-Mobility Components (PEM) entwickelte Formel, wonach sich Industrialisierungsprozesse in der Hälfte der Zeit zu einem Zehntel der Kosten realisieren lassen; ROE = T/2 + I/1 0
Return on Engineering, the formula developed by RWTH at Aachen University's Department of Production Engineering of E-Mobility Components (PEM), according to which industrialization processes can be realized in half the time and at a tenth of the cost; RoE = T /2 + I/1 0

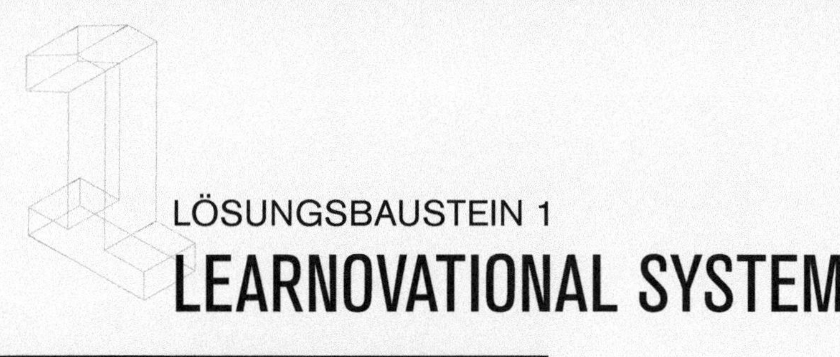

LEARNOVATIONAL SYSTEM

BUILDING BLOCK 1

LEARNOVATIONAL SYSTEM

Am Anfang steht das frühe Scheitern. Und das ist gut. Das Scheitern hat im Rahmen der RoE-Methodik nicht den Ruch der Niederlage oder des Versagens. Entscheidend ist, dass wir Scheitern als Chance begreifen. Es gilt, den Gedanken zu akzeptieren und zu kultivieren, dass Scheitern etwas Positives hat, wenn man daraus kurzfristig Lernimpulse zieht. Handeln gemäß dem „Learnovational System" bedeutet, Innovatives zu schaffen, indem man sich lernend weiterentwickelt. Die Kernaussage lautet: Wir müssen Innovations- und Lernprozess parallelisieren. Der Lösungsbaustein 1 besteht aus 3 Lösungsprinzipien: Hardware Engineering, Flexible PLM-Lösungen, RoE-Kultur.

We had our early fails at the start. And that's good. Within RoE methodology, a fail doesn't carry the taint of defeat or failure. What's crucial is that we see a fail as an opportunity. You've got to accept and cultivate the idea that a fail has a positive aspect if you're able to get little impulses to learn out of it. Acting in accordance with the "Learnovational System" means creating innovation by evolving your learning. The key message is: We need to parallelize innovation and the learning process. Building block 1 consists of 3 principles: Hardware Engineering, Flexible PLM solutions and RoE culture.

LÖSUNGSPRINZIP 1.1.

Hardware Engineering

Grundsätzlich gilt: Ideen müssen anfassbar gemacht werden. Eine Idee ist wertlos, wenn sie nicht in die Umsetzung geht. Um in die Umsetzung gehen zu können, müssen Erfahrungen gesammelt werden. Um Erfahrungen sammeln zu können und daraus Erkenntnisse zu gewinnen, ist Hardware zum Ausprobieren und Testen nötig. Erinnern Sie sich an die Marshmallow-Challenge (Seite 22 ff.)? Man nehme: ungekochte Spaghetti, einen Bindfaden, Klebeband und ein Marshmallow und dann heißt es: Loslegen, ausprobieren, rantasten, Erfahrungen sammeln, Erkenntnisse gewinnen.

PRINCIPLE 1.1.

Hardware engineering

Basically, ideas need to be made tangible. And any idea is worthless if it never makes it to implementation. And to get to implementation, you need to gain experience. And to be able to gain experience and to draw lessons from it you need hardware to test things out. Remember the Marshmallow Challenge (page 22 ff.)? You take uncooked spaghetti, a piece of string, tape and a marshmallow and then get started, try things out, make adjustments, gain experience and gain knowledge.

If you follow the basic principle of trying things out, it is only logical to extend this principle to vehicle development. The hardware on which

Folgt man dem Grundprinzip des Ausprobierens, ist es nur konsequent, dieses Prinzip auch auf die Fahrzeugentwicklung zu übertragen. Die Hardware, an der man seine Erkenntnisse Schritt für Schritt lernend gewinnt, heißt Primotyp. Der Primotyp ist alles andere als ein typischer Prototyp und trotzdem exemplarisch.

Im Gegensatz zum Prototyp ist der Primotyp nicht nur ein auf das Wesentliche reduziertes Versuchsmodell eines späteren Serienprodukts, sondern ein auf Prüfzwecke hin entwickeltes Funktionsmuster im agilen Entwicklungsprozess. Da der Primotyp im Gegensatz zum Prototyp weit einfacher erstellt werden kann, macht er ein schnelleres Ausprobieren von denkbaren Lösungspfaden möglich; wobei das frühe Scheitern bewusst eingeschlossen, ja, gewünscht ist. Dieser Ansatz ist zum Beispiel auch aus der Softwareentwicklung bekannt. Hier hat man verstanden, dass eine Softwareroutine umgehend sichtbare Fehler ausgibt und so auf Konzeptschwächen hinweist. Eine robuste Absicherung dieser Konzeptfehler macht sich langfristig bezahlt.

Folgende Eigenschaften kennzeichnen den Primotyp und grenzen ihn so vom Prototyp ab:

Erkenntnisorientierung

Das Ausprobieren und Lernen hat gegenüber dem Validieren eines Ansatzes (im Sinne Testung/Erprobung) Priorität.

we gain knowledge in a step-by-step process is called a primotype. The primotype is anything but a typical prototype, and yet at the same time it's exemplary.

Unlike the prototype, the primotype is not just a stripped-down test model of a product for ultimate series production, but also a functional model developed for testing purposes in an agile development process. Since unlike the prototype the primotype can be created far more easily, it makes it possible to try out possible solution paths more quickly. In this process, early fails are deliberately part of the process, and actually desired. This approach is also known in the field of software development for instance. Software developers saw that a software routine immediately outputs visible errors thereby pointing out concept weaknesses. A robust protection from such concept errors pays off in the long run.

The following qualities characterize the primotype and distinguish it from the prototype:

Knowledge orientation

Trying out and learning has priority over validating an approach (in the sense of testing/running trials). A professional chef almost always tries out a new dish first: he cooks the dish before he writes the recipe down. And it's the same with vehicles. Example: Design elements can only be constructed iteratively through a large number of prototypes, in

Ein Profikoch probiert ein neues Gericht zunächst immer aus; er kocht das Gericht, bevor er das Rezept aufschreibt.

Und so verhält es sich auch im Fahrzeug. Beispiel: Designelemente können nur über eine Vielzahl von Prototypen iterativ konstruiert werden, also durch schrittweise Annäherung an die exakte Lösung. Zu diesen Designelementen gehören etwa Displays im Fahrzeug, deren Sichtbarkeit unter verschiedenen Lichtverhältnissen sichergestellt sein muss.

Schnittstellenorientierung

Der Prototyp eines Gesamtfahrzeugs entsteht immer relativ spät im Entwicklungsprozess. Innerhalb der herkömmlichen Organisationsstruktur sind die einzelnen Module bis zu diesem Zeitpunkt strikt getrennt. Probleme an den Schnittstellen sind daher häufig und werden erst spät erkannt. Ein Primotyp ist immer ein möglichst vollständiges Versuchsobjekt, so können Probleme an den Schnittstellen früh erkannt und behoben werden – auch wenn die einzelnen Module noch deutlich vom Serienstand abweichen können.

Funktionsorientierung

Dem Prinzip des Ausprobierens zu folgen, heißt, nicht den Prototyp für alles bauen sondern den Primotyp für eine jeweils spezifische Fragestellung.

Beispiel: ein Versuchsfahrzeug für Ergonomieuntersuchungen. Während für den Primotyp

other words, through a gradual approach towards the exact solution. These design elements include things like displays in the vehicle, whose visibility under various lighting conditions has to be ensured.

Interface orientation

The creation of the prototype of a complete vehicle always comes relatively late in the development process. The individual modules are strictly separated up to this point in the conventional organizational structure. Problems with the interfaces are therefore frequent and are identified late. A primotype is always a test object that's as near to complete as possible so that problems at the interfaces can be detected and corrected early – even if the individual modules may still differ significantly from the series standard.

Functional orientation

Following the principle of trial and error means not building a prototype for everything, but instead a primotype for each specific problem.

Example: A test vehicle for ergonomic tests. Whereas wood or cardboard can be used as materials for the primotype, the prototype needs to be quite close to a series model, which can mean much greater expense.

Holz oder Pappe als Werkstoffe genutzt werden können, muss der Prototyp schon nah an der Serie sein, was einen weit höheren Aufwand bedeutet.

Kommunikationsorientierung

Einfach gesagt: Mit dem Primotyp ist immer sichtbar, worüber man redet. Man kann mit dem Finger drauf zeigen, worum es geht. Der Primotyp als Objekt wird so zur Übersetzungs- und Transparenzplattform, die allen Beteiligten stets den aktuellen Entwicklungsstand vor Augen führt. Wenn eine Zusammenarbeit in dezentralen Netzwerken funktionieren soll, wird das extrem wichtig.

Der Primotypenbau – wertvolle Erkenntnisse aus der StreetScooter-Entwicklung

▶ Möglichst einfache, kostengünstige Technologien und Methoden verwenden, etwa gedruckte Werkzeuge oder Lean-Methoden wie das Cardboard Engineering (Modellieren und Simulierung mithilfe von Kartonagen). Zudem starke Fokussierung auf die zu entwickelnde Funktion legen [siehe Funktionsorientierung].

▶ Marketingeffekte dual nutzen. Nicht nur externe Kunden finden neue innovative Projekte spannend, sondern auch die eigenen Mitarbeiter. So entsteht eine hohe Identifikation mit dem Projekt.

Communication orientation

Simply put: With the primotype you can always see what you're talking about. You can point your finger right at it. The primotype as an object thus becomes the translation and transparency platform on which all those involved can see the current state of the development. This becomes extremely important when collaboration in decentralized networks is to function properly.

Building the primotype – valuable insights from the development of StreetScooter

▶ Use the simplest possible, most cost-effective technologies and methods, such as printed tools or lean methods such as cardboard engineering (modeling and simulation using cardboard). Moreover, putting a strong focus on the function to be developed (see functional orientation).

▶ Use marketing effects dually. It's not just external customers who find new innovative projects thrilling – their own employees do too. The result is a high level of identification with the project.

▶ Hardware engineering cannot replace software engineering. Crash tests for example – these must be simulated beforehand.

► Hardware Engineering kann Software Engineering nicht ersetzen. Beispiel Crashtests – diese
müssen vorher simuliert werden.

► Es liegt in der Natur des Ausprobierens, dass
während des Entwicklungsprozesses viele
Erkenntnisse gesammelt werden. Jede Änderung
muss systematisch aufgenommen, kommuni-
ziert und umgesetzt werden. Eine Methode, die
sich anbietet, den Kommunikationsfluss hierar-
chieübergreifend zu verbessern, ist hier das
Shopfloor-Management.

► Absolut notwendig ist eine positive Fehlerkul-
tur. Änderungen darf man nicht als lästig, sondern muss sie als Optimierung empfinden.

Grundsätzliches Fazit:

Vergleicht man das aktuelle StreetScooter-Mo-
dell namens WORK (siehe www.streetscooter.eu/
modelle/work) mit dem ersten Primotyp, fällt auf,
dass sich im Verlauf der Entwicklung vieles geän-
dert hat. Und das ist gut so. Wir haben früh aus
den Erfahrungen gelernt und so konnte aus dem
Ur-Modell des StreetScooters Schritt für Schritt der
gelbe StreetScooter der Deutschen Post werden,
wie man ihn live aus der täglichen Zustellung oder
aus den Medien kennt.

► It's in the nature of trial and error that many
findings are collected during the development process. Any change has to be syste-
matically recorded, communicated and implemented. One method that lends itself to
improving the flow of communication across
the hierarchy here is shop floor management.

► A positive error culture is an absolute must.
Changes should not be seen as a nuisance,
they should be seen as optimization.

Basic conclusion:

If you compare the current StreetScooter model
called WORK (see www.streetscooter.eu/modelle/
work) with the first primotype, you can see how
much has changed in the course of development.
And that is a good thing. We learned from experience early on, and that's what allowed the original model of the StreetScooter to gradually become
the yellow StreetScooter of the Deutsche Post, as we
know it from actual daily delivery in the street or
from the media.

LÖSUNGSPRINZIP 1.2.

Flexible PLM-Lösungen

Lernen im schrittweisen Entwicklungsprozess führt ständig zu Veränderungen. Hier ist ein ebenso ständiger Austausch zwischen den Partnern erforderlich, vor allem bei Netzwerkinnovationen mit dezentraler Struktur wie bei der Entwicklung des StreetScooters, bei dem bis zu 80 ausgewählte Partnerunternehmen beteiligt waren. Eine koordinierte Zusammenarbeit war nur mit einem zentralen IT-Tool möglich und dem Austausch über den aktuellen Entwicklungsstand.

Innovative Informations- und Kommunikationstechnologie-Lösungen (kurz I&K-Lösungen) sind entscheidend für eine funktionierende und effiziente Zusammenarbeit im Netzwerk. Dieser Überzeugung sind laut einer Studie[20] 68,8 % der befragten OEM[21] und Zulieferer. Sie sehen in der Entwicklung neuer I&K-Lösungen den entscheidenden Hebel für ein erfolgreiches Management komplexer Netzwerke. Hier kommen PLM-Lösungen ins Spiel. PLM steht für Product Lifecycle Management, womit das Managen aller Daten und Informationen eines Produkts über den gesamten Lebenszyklus gemeint ist. Das Produkt erhält dabei über den Lebenszyklus einen digitalen „Schatten", in dem alle Daten und Änderungen gespeichert werden, auf den dann verschiedene Programme zugreifen können. Es existiert also eine zentrale Datenquelle (Single Source of Truth), mit der alle

PRINCIPLE 1.2.

Flexible PLM solutions

Learning in the gradual development process constantly leads to changes. Here you need an equally constant exchange between the partners, especially in network innovations with a decentralized structure as we had with the development of the StreetScooter, in which up to 80 selected partner companies were involved. A coordinated collaboration was only possible with a central IT tool, and with exchanges about the current state of development.

Innovative ICT solutions (I&C solutions) are critical for a functioning and efficient collaboration within the network. According to a study[20], 68.8 % of the OEM[21] and suppliers surveyed agreed with this. They see in the development of new I&C solutions the key levers for successful management of complex networks. This is where PLM solutions come into play. PLM stands for Product Lifecycle Management, which refers to managing all data and information of a product throughout the entire lifecycle. The product thereby receives a digital "shadow" over its lifecycle, in which all data and changes are saved, which various programs can then access. There is also a central data source (Single Source of Truth), to which all other systems (such as CAD or CRM) are linked. In this way, approvals of design changes in the network (i. e., company-wide) are possible, for instance.

20 Quelle: Kurzstudie RWTH Aachen, 2012
Source: Short study RWTH Aachen, 2012

21 OEM, englisch Original Equipment Manufacturer; übersetzt Originalausrüstungshersteller. Hier sind darunter die etablierten Automobilhersteller zu verstehen.
OEM, Original Equipment Manufacturer. These include the established automotive manufacturers.

anderen Systeme (wie CAD oder CRM) verknüpft sind. Auf diese Weise sind zum Beispiel Freigaben von Konstruktionsänderungen im Netzwerk, also unternehmensübergreifend, möglich.

Netzwerk-Kommunikation – wertvolle Erkenntnisse aus der StreetScooter-Entwicklung

► Organisationsstrukturen: klare Zuständigkeiten im Netzwerk; klare Entscheidungsprozesse; Transparenz über Partner-Kompetenzen und –Verantwortlichkeiten; definierte Ansprechpartner bei allen Unternehmen und für jede Funktion.

► Datenaustausch: Alle im Markt erhältlichen CAD-Systeme waren im StreetScooter-Netzwerk enthalten; Datenaustausch und Datenintegration waren nur über das PLM-System als sichere Daten-Plattform möglich.

► Schnittstellenmanagement: frühe virtuelle Integration von Komponenten; Transparenz über Auswirkungen von Änderungen im Produkt; klare Zuordnung der Zuständigkeiten; Schnittstellen bestanden zwischen Partnern, die viele Kilometer voneinander entfernt waren.

► Fortschrittscontrolling: dank zentralen Datenstandes einfaches Fortschrittscontrolling; Transparenz über den jeweils relevanten Datenstand durch eindeutige Rechtevergabe für jeden Nutzer; durch die räumliche

Network communication – valuable insights from the development of StreetScooter

► Organizational structures: clear responsibilities in the network; clear decision-making processes; transparency about partner competencies and responsibilities; defined contacts for all companies and for each function.

► Data exchange: all CAD systems available on the market were included in the StreetScooter network; data exchange and data integration was possible only via the PLM system as a secure data platform.

► Interface management: early virtual integration of components; transparency of the effects of changes in the product; clear assignment of responsibilities; interfaces existed between partners who were many miles apart.

► Progress controlling: easy progress controlling thanks to central data status; transparency of the relevant data status through clear assignment of rights for each user; virtual proximity and transparency is needed, given the physical distance between partners.

► Change management: all changes go through an approval process in the network; affected network partners are informed; the change

Distanz zwischen Partnern ist eine virtuelle Nähe und Transparenz notwendig.

► Änderungsmanagement: Jede Änderung durchläuft einen Freigabeprozess im Netzwerk; betroffene Netzwerkpartner werden informiert; die Änderungshistorie ist stets transparent, was bei agilem Vorgehen mit vielen Änderungen wichtig ist.

LÖSUNGSPRINZIP 1.3.

RoE-Kultur

Lernen in einem Fluss kontinuierlicher Veränderungen heißt lernen aus Fehlern. Der Umgang mit Fehlern erfordert eine entsprechende Unternehmenskultur, die bei uns in Deutschland grundsätzlich wenig ausgeprägt ist. Der deutsche Ingenieur ist durch seine klassische Ausbildung auf Perfektion trainiert. Das Entwickeln von einfachen und pragmatischen Lösungen, wie sie für einen Primotyp nötig sind, ist nicht Teil seiner Ausbildung. Den positiven, sprich konstruktiven Umgang mit Fehlern muss er sich erst noch aneignen.

Entwickeln ohne Spezifikationen?

Klingt provokativ, zumindest ungewöhnlich. Für viele Partner war das bei der Entwicklung des StreetScooters zunächst eine neue Erfahrung.

history is always transparent, which is important in an agile approach with many changes.

PRINCIPLE 1.3.

RoE culture

Learning in a flow of continuous changes means learning from mistakes. Dealing with errors demands a corresponding corporate culture, one that is basically underdeveloped in Germany. The classical education German engineers receive trains them for perfection. Developing simple and pragmatic solutions such as are needed for a primotype is not part of their training. Knowing how to handle errors in a positive way – i.e., constructively – is something they have yet to acquire.

Developing without specifications?

Sounds provocative, or unusual at the least. This was initially a new experience for many partners in the development of the StreetScooter.

They were used to being given specifications dictated to them in a top-down way. Now they were expected to determine the specifications together with the other partners. And for a good reason: We wanted to start with a white sheet of

Sie waren es gewohnt, Spezifikationen top-down vorgegeben zu bekommen. Jetzt sollten sie die Spezifikationen gemeinsam mit den anderen Partnern festlegen. Und das aus gutem Grund: Wir wollten auf dem weißen Blatt Papier beginnen und frei von Restriktionen ein kostengünstiges Elektroauto entwickeln. Ganz entscheidend ist dazu die Kommunikation entlang der gesamten Wertschöpfungskette gewesen. Nur so ließen sich Kostenpotenziale gemeinsam heben.

Neben der Entwicklung technischer Lösungen musste auch die neue Art der Zusammenarbeit trainiert werden. Die Analogie zu diesem Prozess heißt Lean Production[22]. Aber auch bei dieser Methode reicht es nicht, einige Methoden anzuwenden und dann ist das Unternehmen lean. Vielmehr muss ein grundsätzlicher kultureller Wandel stattfinden, der durch verschiedene Faktoren geprägt wird und ein hohes Investment des Unternehmens erfordert, das sich aber auszahlt. Belegen lässt sich das durch eine Studie[23], die an der RWTH Aachen durchgeführt wurde. Das Ergebnis: Unternehmen, die mehr Mitarbeiter für ihr Produktionssystem zur Einführung von Lean Production abstellen und die mehr in das Thema investieren, sind auch überdurchschnittlich erfolgreich.

Auf dem Feld der Entwicklung gibt es bisher nichts Vergleichbares. Beispiel: Simultaneous Engineering[24] (kurz SE). Zur Umsetzung von Simultaneous Engineering werden zwar auch SE-Teams eingeführt, verbunden damit ist aber kein vergleichbarer Management-Prozess wie bei Lean Production. Zur Einführung und Anwendung von Lean

paper and develop a low-cost electric car free of restrictions. Communication along the entire value chain was absolutely mission critical here. It was the only way to jointly raise cost potentials.

In addition to the development of technical solutions, the new type of collaboration needed to be trained. The analogy for this process is called Lean Production[22]. But even with this method, it's not just about applying a few methods and then the company is suddenly lean. Rather, a fundamental cultural change needs to take place, which is influenced by various factors and demands a high investment of the company, but one that pays off. Proof of this can be obtained from a study[23] conducted at the RWTH Aachen. The result: Companies that dedicate more staff to their production system for the introduction of lean production and invest more in the topic also prove to be companies with above-average success.

There has thus far been nothing comparable in the field of development however. Example: Simultaneous Engineering[24] (SE). In order to implement simultaneous engineering, SE teams are actually introduced, but then there is no comparable management process connected with it as there is with lean production. The introduction and application of lean production involves benchmarks, training programs, specially appointed employees, etc. Consequently, the degree of implementation of lean production is many times higher than in SE. It should however be said by way of precaution that processes never apply exactly 1:1. A formal custom implementation in the company is always a critical

22 Lean Production oder Lean Manufacturing, übersetzt Schlanke Produktion, darunter versteht man den sowohl sparsamen als auch zeiteffizienten Einsatz der Produktionsfaktoren Betriebsmittel, Personal, Werkstoffe, Planung und Organisation im Rahmen aller Unternehmensaktivitäten.
Lean Production or Lean Manufacturing refers to both economical and time-efficient use of production factors of resources, personnel, materials, planning, and organization, within the framework of all company activities.

23 Quelle: Studie „Konsortial-Benchmarking – Production Systems", Studienergebnisse & Projekthighlights, RWTH Aachen, 2015
Source: Study "Konsortial-Benchmarking – Production Systems" Studienergebnisse & Projekthighlights, RWTH Aachen, 2015

24 Simultaneous Engineering, Methode zur Verkürzung der Produktentwicklung mittels paralleler Initiierung der notwendigen Entwicklungsarbeiten. Um die Entwicklung zu beschleunigen und zu optimieren, können in den simultanen Prozess auch die Lieferanten einbezogen werden.
Simultaneous Engineering, a method for shortening product development by using parallel initiation of the necessary development work. The suppliers may be involved in the simultaneous process in order to accelerate and optimize development.

Production gehören Benchmarks, Ausbildungsprogramme, speziell abgestellte Mitarbeiter u.v.m. Entsprechend ist der Umsetzungsgrad von Lean Production um ein Vielfaches höher als bei SE. Vorsorglich ist jedoch zu sagen, dass sich Prozesse nie 1:1 übertragen lassen. Entscheidend ist immer eine formal individuelle Umsetzung im Unternehmen. Es lässt sich aber klar festhalten, dass Unternehmen, die an diesem Kulturwandel arbeiten und sich individuell entwickeln, erfolgreicher sind, als die, die nur kopieren – auch das zeigt die o.g. Studie der RWTH. Wichtig ist, dass Veränderungen im Industrialisierungsprozess durch einen kulturellen Wandel begleitet werden müssen.

Unternehmenskulturelle RoE-Entwicklung – wertvolle Erkenntnisse aus der StreetScooter-Entwicklung

► Strategische Initiierung

► Entwicklung einer Vision

► Top-down-Umsetzung

► Einführung und ganzheitliche Mitarbeitereinbindung

► Mitarbeiter zu „Überzeugungstätern" machen

► Unternehmertum bilden

► Nachhaltigkeit betonen

► Glaubwürdigkeit leben

factor. But it can be clearly said that companies that work on this cultural change and which evolve individually are more successful than those which only copy – and the abovementioned RWTH study shows this as well. It is important that changes in the industrialization process be accompanied by a corresponding cultural change.

Organizational cultural RoE development – valuable insights from the development of StreetScooter

► Strategic initiation

► Developing a vision

► Top-down implementation

► Introduction and holistic employee involvement

► Turning employees into "people of conviction"

► Instituting entrepreneurship

► Stressing sustainability

► Living credibility

Grundsätzliches Fazit:

Bei der Entwicklung des StreetScooters hat sich gezeigt, dass es besonders wichtig ist, den gewollten kulturellen Wandel klar zum Ausdruck zu bringen, um so einen sinnstiftenden Kern für Identität zu schaffen. Das Ergebnis: Alle ziehen an einem Strang. Die Grundphilosophie des RoE ist allen verständlich und wird bewusst gelebt. Auch das Personal Recruiting ist auf Bewerber ausgerichtet, die diese Kultur aus Überzeugung leben und auch nach außen transportieren wollen.

Basic conclusion:

In developing the StreetScooter it was shown that it is particularly important to clearly articulate the desired cultural change in order to provide a meaningful core for identity. The result: Everyone pulling together. The basic philosophy of the RoE is understandable to all and consciously lived. Even staff recruitment is oriented towards candidates who live this culture by conviction and who wish to transport it to the outside world.

Customer Engineering erfordert ein
klares Priorisieren der Anforderungen,
um Over-Engineering zu vermeiden
Customer Engineering requires a clear
prioritization of the requirements in
order to avoid over-engineering

CUSTOMER ENGINEERING

BUILDING BLOCK 2

CUSTOMER ENGINEERING

Im vorigen Kapitel (Lösungsbaustein 1 „Learnovational System") wurden die Vorteile des frühen Scheiterns erklärt. Der Modus lautet: Scheitern ist nützlich, wenn man daraus Erkenntnisse zieht und sie umsetzt. Fehler machen zu dürfen, weil man daraus lernen kann, ist ein wesentliches Merkmal der RoE-Methodik. Ein weiteres Merkmal dieser Methodik ist die Idee, ein Produkt zu entwickeln (wie den StreetScooter), das individuell auf die Bedürfnisse und Anforderungen des Kunden eingeht. Nicht Produktvielfalt bei gleichzeitig größtmöglicher Variantenbreite steht im Mittelpunkt, sondern die individuelle Anwendung; entwickelt und gebaut wird das, was der Kunde wirklich braucht. Mit anderen Worten: Die Praxis gibt die Spezifikation vor. Die Kernaussage lautet: Nicht noch mehr Varianten und ein immer breiteres Produktprogramm anbieten, sondern effiziente Individualisierung ermöglichen. Der Lösungsbaustein 2 besteht aus 3 Lösungsprinzipien: Kundenintegration, Spezifikationsmanagement, Risikomanagement.

In the preceding chapter (Building Block 1 "Learnovational System") the benefits of early failure were explained. According to the mode: Failure is useful if you draw insights from it and implement them. Being allowed to make mistakes, because you can learn from them is an essential feature of the RoE methodology. Another feature of this methodology is the idea of developing a product (such as the StreetScooter) that addresses the individual needs and requirements of the customer. The focus then is not on product variety with simultaneous maximum variation width, but on the individual application instead, i. e., developing and building what the customer really needs. In other words: Actual use dictates the specifications. The key message is: No longer offer even more variants and an ever-wider range of products, but enable efficient individualization instead. Building block 2 consists of 3 principles: Customer integration, specification management and risk management.

LÖSUNGSPRINZIP 2.1.

Kundenintegration

Schon heute entwickeln bis zu 40 % der Fahrzeug-Kunden ihre Produkte weiter und kreieren eigene Derivate[25] – vom Nachrüsten der Freisprecheinrichtung bis hin zu individuell angepassten Flottenfahrzeugen. Nach dem Motto: Nicht

PRINCIPLE 2.1.

Customer integration

Even today, up to 40 % of vehicle customers further develop their products and create their own derivates[25] – from retrofitting the speakerphone to customized fleet vehicles. Following the motto: Don't just complain: change things yourself. This mindset is why the topic of innovation is seeing an increasing democratization. Result: If customers

25 Derivat, darunter versteht man eine Produktvariante, die von einem Basis-Produkt abgeleitet wird
Derivate: this refers to a product variant that is derived from a basic product

meckern, sondern selber ändern, erfährt das Thema Innovation so eine zunehmende Demokratisierung. Fazit: Wenn Kunden schon selbst hergehen und Produkte oder Features an eigene Bedürfnisse anpassen oder weiterentwickeln, warum sie dann nicht gleich in den ursprünglichen Entwicklungsprozess einbeziehen?

Gemeinsam sind wir schlauer – aber auch effizienter

Effizient individualisieren heißt, Produkte nicht für, sondern mit dem Kunden entwickeln. Der Vorteil: Kundenindividuelle Produkte sind deutlich profitabler, wenn sie effizient individualisiert wurden. Schon ein Derivat bietet ein deutlich höheres Preispotenzial als ein Standardprodukt, effizient individualisierte Lösungen sind zusätzlich attraktiv.

Bei der Kundenintegration können zwei Wege beschritten werden. Entweder das Produkt wird kundenindividuell produziert (Beispiel: der NikeiD-Schuh[26]) oder der Kunde kann nach dem Kauf noch Änderungen vornehmen (Beispiel: Apps fürs Handy).

Fahrzeugentwicklung und Kundenintegration

Kundenintegration lässt sich auf die Fahrzeugentwicklung übertragen, stellt aber eine besondere Herausforderung dar. Die Gründe dafür sind u.a.:

are already alongside you, customizing or further developing products or features according to their own needs, why not simply include them in the original development process itself?

Together we are smarter – but also more efficient

Efficient customization means developing products not for the customer, but together with him. The advantage: Customized products are much more profitable if they have been individualized efficiently. A derivate already offers a significantly higher price potential than a standard product: efficiently customized solutions are furthermore more attractive.

Two paths to customer integration can be taken. Either the product is tailor produced (e.g., the NikeiD shoe[26]) or the customer can make changes after the purchase (e.g., apps for cell phones).

Vehicle development and customer integration

Customer integration can be applied to vehicle development, but it presents a special challenge. Some of the reasons for this are:

26 NikeiD ist ein personalisierter Schuh, den der Kunde selbst entwirft; mittels 3D-Designer kann der Wunschschuh in Farbe und Material selbst gestaltet und mit einer eigenen ID versehen werden. NikeiD is a customized shoe, which customers design themselves; using 3D designer, the desired shoe's color and materials can be set and the shoe itself is given its own ID.

- ► Eine Veränderung des Produkts kann sicherheitskritisch sein, ggfs. muss eine neue Straßenzulassung erfolgen.

- ► Es besteht eine hohe technische Abhängigkeit zwischen den Modulen.

- ► Ein Fahrzeug ist Hardware, Hardware ist im Gegensatz zur Software schwer zu individualisieren.

- ► Produktionsprozess und Lieferkette sind komplex, daher schwer effizient und schnell zu individualisieren.

Additive Manufacturing[27] als Schlüssel für effiziente Individualisierung

Bei der individuellen Produktion eines Fahrzeugs bietet das Additive Manufacturing ein sehr großes Potenzial. Eine Individualisierung von einzelnen Komponenten ist vielseitig denkbar, etwa wenn es um eine perfekt angepasste Ergonomie geht. In Zukunft ist sogar die Produktion von ganzen Fahrzeugen im Additive Manufacturing denkbar, das heißt, Fahrzeuge werden im Shop ausgedruckt, erste Versuche dazu gibt es. Hinsichtlich offener Schnittstellen sind besonders im Fahrzeuginnenbereich Potenziale vorhanden. Bis dato kann der Kunde im Interieur keine echten Änderungen vornehmen. Möglich ist nur noch Tuning, aber kein Nachrüsten von Modulen oder Komponenten.

Bei der Entwicklung und Weiterentwicklung des StreetScooters war Kundenintegration

- ► A modification of the product can be critical to security, for instance, a new street legal approval may apply.

- ► There is a high technical dependency between the modules.

- ► A vehicle is hardware, and hardware, unlike software, is hard to customize.

- ► Production process and supply chain are complex, thus making it hard to individualize them efficiently and quickly.

Additive manufacturing[27] as the key to efficient individualization

In the individual production of a vehicle, additive manufacturing offers enormous potential. An individualization of individual components is conceivable in many ways, for instance when it comes to perfectly adapted ergonomics. In the future, even the production of complete vehicles in additive manufacturing is conceivable; vehicles will be printed at the store, something that has already been tested. With regard to open interfaces, there is potential especially in the passenger area of the vehicle. To date, customers cannot make any real changes to the interiors. The only option today is tuning, but no retrofitting of modules or components.

With the development and evolution of the StreetScooter, customer integration was self-evident; above all this meant the early involvement

27 Additive Manufacturing, übersetzt Generative Fertigung bzw. Additive Fertigung (zum Beispiel das 3D-Druck-Verfahren), ist eine umfassende Bezeichnung für Verfahren zur schnellen und kostengünstigen Fertigung von Modellen, Mustern, Prototypen, Werkzeugen und Endprodukten. Die Fertigung erfolgt direkt auf der Basis rechnerinterner Datenmodelle. Additive manufacturing (e. g., the 3D printing process) is a comprehensive term for methods for the rapid and cost-effective manufacturing of models, patterns, prototypes, tools and finished products. The manufacturing is carried out directly on the basis of computer-aided data models.

selbstverständlich; allem voran ist damit die frühe Einbeziehung des Deutsche-Post-Kunden in die Entwicklung gemeint. Der Kunde hat das Produkt bereits zu einem frühen Zeitpunkt definiert; nicht erst am Ende der Vorserie, sondern maßgeblich bereits bei der Entwicklung der Spezifikationen. Um die Anforderungen aufzunehmen, wurden viele Primotypen genutzt, sodass der Kunde sich das Ergebnis immer vorstellen und es ausprobieren konnte.

Kundenintegration – wertvolle Erkenntnisse aus der StreetScooter-Entwicklung

Die Struktur des StreetScooters hat eine effiziente Individualisierung ermöglicht:

► die Baukastenstruktur schafft hohe Flexibilität

► auf Flexibilität setzen, um kurze Entwicklungsprozesse möglich zu machen

► ebenso variable und flexible Technologien, die mit wenig Investitionen umsetzbar sind, nutzen

► durch Individualisierung spart der Kunde nicht nur Geld, auch die Mitarbeiter sind zufriedener und identifizieren sich mit dem Fahrzeug

► im Falle Deutsche Post wurden alle Hierarchieebenen im Unternehmen abgefragt

► schrittweise vorgehen, Veränderungen bewerten

► selbstkritisch sein, Optimierung suchen und umsetzen

of the Deutsche Post into the development. The customer had already defined the product at an early stage; not only at the end of pre-production, but mainly during the development of the specifications. Many primotypes were used to accommodate the requirements, so that the customer could always imagine the result and test it out.

Customer integration – valuable insights from the development of StreetScooter

The structure of the StreetScooter has an efficient individualization so that:

► the modular structure provides high flexibility

► it is set up for flexibility in order to make short development processes possible

► it uses comparably variable and flexible technologies that can be implemented with little investment

► through individualization, the customer not only saves money but the employees are more satisfied and identify with the vehicle

► in the case of Deutsche Post all hierarchical levels in the company were queried for input

► there is a gradual approach, changes can be assessed

► there is room for being self-critical, optimization can be sought out and implemented

Grundsätzliches Fazit:

Blickt man auf den StreetScooter, lassen sich viele Beispiele für eine Individualisierung vorlegen. Etwa der in drei Komponenten geteilte Schweller. Die einzelnen Teile lassen sich separat herausnehmen und austauschen. Das ist günstiger, als die komplette Komponente auszutauschen oder zu reparieren. Ein anderes Beispiel ist die seitliche Laderaumtür, deren Robustheit speziell auf bis zu 250 tägliche Bewegungen (Öffnen, Schließen) ausgelegt ist. Weitere Beispiele für Individualisierungen lassen sich im Ergonomiebereich oder im Laderaum finden.

LÖSUNGSPRINZIP 2.2.

Spezifikationsmanagement

Customer Engineering erfordert ein klares Priorisieren von Anforderungen: Was ist Standard, was ist individuell? Der Kernsatz lautet: Auf Basisanforderungen beim Standard fokussieren und dann effizient individualisieren. Dabei geht es darum, jegliches Over-Engineering zu vermeiden und das Produkt derart auf die spezifischen Kundenbedürfnisse zuzuschneiden, dass diese genau erfüllt, aber nicht übererfüllt werden, denn Übererfüllung verursacht hohe Kosten.

Basic conclusion:

If you look at the StreetScooter, many examples of individualization present themselves. For instance, in the rocker panel, divided into three components. The individual parts can be removed separately and replaced. This is cheaper than replacing the entire component or repairing it. Another example is the side cargo door, whose robustness was specially designed to handle up to 250 movements a day (opening and closing). Other examples of customization can be found in the area of ergonomics or in the cargo area.

PRINCIPLE 2.2.

Specification management

Customer engineering requires a clear prioritization of requirements: what is standard, what is customized? The key here is to focus on basic requirements for the standard and then customize efficiently. The aim is to avoid any over-engineering and to tailor the product to specific customer needs so these are exactly met, but not surpassed because overachieving can result in high costs.

Spezifikationen auf dem Prüfstand

In der Praxis ist es häufig so, dass Spezifikationen über Produktgenerationen entwickelt und gar nicht mehr hinterfragt werden. Ungesteuertes Priorisieren muss jedoch unbedingt vermieden werden. Dem gezielten Managen von Spezifikationen kommt damit eine besondere Bedeutung zu.

Nachteile fixer Spezifikationen

Nach dem Status quo werden Spezifikationen zu Beginn festgesetzt und nur noch angepasst, wenn sie nicht erfüllbar sind. Ein Optimierungsprozess findet nicht statt.

► Spezifikationen werden häufig einfach durch interne Teams definiert und nicht aus Kundensicht hinterfragt.

► Kunden können zu Beginn nur schwer befragt werden, da das Produkt noch nicht entwickelt ist.

► Werden Kunden spät im Prozess befragt, können sie zwar Feedback geben, aber es können keine Änderungen mehr vorgenommen werden.

► Viele Unternehmen sind gar nicht mit dem Endkunden im Kontakt und können diesen somit kaum direkt befragen.

► Spezifikationen sind über Unternehmensgrenzen hinweg meist nicht veränderbar.

► Spezifikationen anzupassen braucht Überzeugungsarbeit. Meist sind dafür erfolgreiche

Putting specifications to the test

In practice, it is often the case that specifications are developed through product generations and no longer questioned. And yet uncontrolled prioritizing has to be avoided at all costs. The targeted management of specifications is thus of particular importance.

Drawbacks of fixed specifications

As per the status quo, specifications are fixed at the beginning and only adjusted when they cannot be fulfilled. There is no optimization process.

► Specifications are often simply defined by internal teams and not questioned from a customer perspective.

► It is very difficult to query customers at the start because the product has not yet been developed.

► If customers are queried late in the process, they can indeed give feedback, but then it's too late to make any changes.

► Many companies actually have no contact with the end users, so they can hardly query them directly.

► Specifications cannot generally be changed across company boundaries.

► Adapting specifications takes some persuasion. As a rule, reference projects are necessary to

Referenzprojekte notwendig; Spezifikationen sind im Sinne einer Optimierung für das Gesamtsystem nicht verhandelbar.

► Die Auswirkungen von Spezifikationen sind meist nicht transparent; es gibt keine klare Kostentransparenz, was kostet oder spart eine Änderung in den Spezifikationen?

Spezifikationsmanagement – wertvolle Erkenntnisse aus der StreetScooter-Entwicklung

► Wettbewerb der Lösungen: Um Kostenpotenziale heben zu können, erhielten die Zulieferer nur ungefähre Spezifikationen. Wichtigste Vorgabe war stets der Zielpreis je Komponente. Im Kern ging es darum, das Know-how und die Kreativität der Lieferanten zu aktivieren, um einen Wettbewerb der Lösungen zu initiieren.

► Auswirkung von Spezifikation auf Kosten: Sämtliche Qualitätsanforderungen wurden über die gesamte Wertschöpfungskette diskutiert und festgelegt. Ein Beispiel ist die Qualität des Kupferdrahts gewesen. Hier wurde in einer Diskussion mit den Partnern auf der gesamten Wertschöpfungskette herausgearbeitet, dass die Qualitätsanforderungen an den Draht um den Faktor 10 reduziert werden können, ohne Qualitätseinbußen im Endprodukt zu erhalten. Mit dem Verständnis von Materiallieferant, Wickelanlagenhersteller und

convince others; specifications are not negotiable in terms of optimization of the overall system.

► The effects of specifications are usually not transparent; there is no clear cost transparency as to what a change in the specifications might cost or save?

Specification management – valuable insights from the development of StreetScooter

► Competition of solutions: In order to raise cost potentials, the suppliers were given only approximate specifications. The main stipulation was always the target price per component. In essence, this was about activating the expertise and creativity of the suppliers in order to initiate a competition of solutions.

► Impact of specification on costs: All quality requirements were discussed and defined across the entire value chain. One example was the quality of the copper wire. Here in a discussion with the partners in the total value chain, it was worked out that the quality requirements for the wire could be reduced by a factor of 10 without sacrificing quality in the final product. With the understanding of material suppliers, winding equipment manufacturers and engine manufacturers, the requirements were coordinated

Motorenhersteller wurden die Anforderungen abgestimmt und letztendlich durch geringere Anforderungen deutliche Kosten eingespart.

▶ Gleichberechtigung der Partner: Gemeinsam mit allen Beteiligten stellten wir uns immer wieder die Frage, wie sich diese oder jene Spezifikation auf die Kosten auswirken würde. Die Frage war für jede Komponente individuell zu beantworten. Entscheidend war in der Diskussion eine Gleichberechtigung der Partner auf den unterschiedlichen Wertschöpfungsstufen.

▶ Bewertung der Kundennutzen: Um Over-Engineering zu vermeiden, wurden die Kunden in den Prozess zur Abstimmung der Spezifikationen integriert. Der Kundennutzen wurde für die einzelnen Komponenten individuell bewertet und damit die Spezifikationen beeinflusst.

▶ Prozess-Anforderung: Es geht nicht nur um Produktspezifikationen, sondern genauso um Anforderungen an die Produktionsprozesse.

Grundsätzliches Fazit:

Die Festlegung der Spezifikationen auf dem berühmten weißen Blatt Papier. Bei null zu beginnen bedeutete Chance und Risiko zugleich. Diese im Falle des StreetScooters gewählte Herangehensweise an das Spezifikationsmanagement erforderte ein radikales Umdenken bei den Zulieferern. Für viele Zulieferer war es völlig ungewohnt, Spezifikationen zwischen Partnern zu verhandeln. Und auch mit offenen Spezifikationen umzugehen war für viele fremd.

and ultimately there were significant cost savings as a result of the lowered requirements.

▶ Equality of the partners: Together with all stakeholders, we constantly asked ourselves the question of how this or that specification would affect the cost itself. The question was to be answered individually for each component. The equality of the partners was critical in the discussions at the various stages of the value chain.

▶ Assessing customer value: To avoid over-engineering, the customers were integrated into the process of gaining consensus on the specifications. The customer benefit was assessed individually for each component, which then influenced the specifications.

▶ Process requirement: It's not just about product specifications, but also about requirements for the production processes.

Basic conclusion:

The defining of the specifications on the famous white sheet of paper. Starting at zero means at once opportunity and risk. In the case of StreetScooter, this chosen approach to the specification management demanded a radical rethinking among suppliers. For many suppliers, negotiating specifications between partners was sailing in uncharted waters. And for many, dealing with open specifications was also quite strange.

Ein Beispiel aus dem Workshop „Karosserieent-
wicklung" macht diese Problematik deutlich.
StreetScooter-Frage: Wie sieht die Karosserie
idealerweise aus, um sie günstig produzieren zu
können? Zulieferer-Antwort: Zeig´ mir die Komponenten, dann können wir die Kosten nennen.
StreetScooter-Einwand: Dann ist es aber schon zu
spät. Bestimmte Rahmenbedingungen sind dann
bereits gesetzt ...

An example from the workshop "Body Development" makes this problem evident. StreetScooter
question: What would the body ideally look like
in order to produce it cheaply? Supplier response:
Show me the components, then we'll quote the
costs. StreetScooter objection: But then that'd be
too late. Certain conditions are already set ...

LÖSUNGSPRINZIP 2.3.

Risikomanagement

Wenn Industrialisierungszeiten verkürzt und
-budgets reduziert werden, wenn Produkte darüber hinaus kundenindividuell sind und keine Stan-
dard-Massenware, wenn das alles gewollt ist, weil
es zur Strategie gehört, muss man akzeptieren,
dass das Risiko des Scheiterns steigt, und zwar
sowohl in technischer als auch in wirtschaftlicher
Hinsicht. Dies betraf unisono die Entwicklung des
StreetScooters. Dem galt es durch ein gezieltes
Risikomanagement entgegenzuwirken.

Technisches Risikomanagement bedeutet nicht nur klassische Absicherungsprozesse

► Testing in Hardware und das Sammeln von
  Erfahrungen reduzieren das Risiko; dem

PRINCIPLE 2.3.

Risk management

If industrialization times are shortened and bud-
gets reduced, if products are furthermore customized and are not standard commodities, if everything is intentional, because it is part of the
strategy, then you have to accept that the risk
of failure increases, both technically and economically. This applied to the development of the
StreetScooter across the board. A targeted risk
management was need to counter it.

Technical risk management means more than simply classic risk protection processes

► Testing in hardware and collecting experiences reduces the risk; the use of primotypes is
  therefore of particular importance.

28 Connectivity, darunter versteht man Integrationslösungen für mobile Geräte im Fahrzeug und zur Vernetzung von Fahrzeugen mit der Außenwelt
Connectivity, this refers to integration solutions for mobile devices in the vehicle and to the networking of vehicles with the outside world

29 Data Analytics, steht für die Untersuchung bestimmter Datenmengen unterschiedlicher Art, um darin Muster, unbekannte Korrelationen oder andere nützliche Informationen zu entdecken
Data analytics, is the investigation of certain amounts of data of different types to discover patterns, unknown correlations and other useful information

30 Disruptive Technologien, eine Innovation, die eine bestehende Technologie, ein etabliertes Produkt oder eine bekannte Dienstleistung möglicherweise vollständig verdrängt
Disruptive technologies, an innovation that may completely supplant an existing technology, an established product or service

31 Lead-User, wird als trendführender Nutzer oder trendführender Kunde verstanden; es sind User, deren Bedürfnisse den Anforderungen des Massenmarktes voraus sind
Lead users, refers to trend leading users or customers; these are users whose needs are far ahead of the requirements of the mass market

Einsatz von Primotypen kommt daher eine besondere Bedeutung zu.

► Neue technische Lösungen im Bereich Connectivity[28] und Data Analytics[29] halfen, automatisiert Daten im Testing und Betrieb aufzunehmen und auswerten zu können und so das Risiko zu senken.

Die Absicherung des wirtschaftlichen Risikos ist vor allem bei disruptiven Technologien[30] interessant

Wirtschaftliches Risiko besteht dann, wenn der Erfolg im Markt nicht vorhersagbar ist. Dies betrifft vor allem disruptive Produkte. Denn wenn es den Markt für das zu entwickelnde Produkt noch gar nicht gibt, kann dieser auch nicht analysiert werden. Dem gilt es entgegenzuwirken, indem in sogenannten Marktexpeditionen das Absatzpotenzial des Produkts bei ausgewählten Kundengruppen untersucht wird. Für die Marktexpedition erforderlich sind drei Schritte:

► Auswahl und Identifikation relevanter Lead-User[31], wichtig ist dabei u. a. eine besondere Technologieaffinität der Personen oder Unternehmen.

► Entwicklung eines kontextspezifischen Designs, auf Lead-User zugeschnittenes Produkt.

► Durchführung und Bewertung intensiv verstehen.

► New technical solutions in the field of connectivity[28] and data analytics[29] helped in recording and evaluating automated data in the area of testing and operation, thereby reducing the risk.

The hedge against economic risk is of particular interest in disruptive technologies[30]

Economic risk exists when the success in the market is not predictable. This particularly applies to disruptive products. Because if there's no market yet for the product to be developed, then by extension, it can't be analyzed either. Accordingly, this needs to be countered by testing the sales potential of the product in selected groups of customers in market expeditions. The market expeditions require three steps:

► Selection and identification of relevant lead users[31], where a special affinity for technology by the people or companies is one key criterion.

► Development of context-specific designs for a product tailored to lead users.

► Intensely understanding implementation and assessment.

Risikomanagement – wertvolle Erkenntnisse aus der StreetScooter-Entwicklung

► StreetScooter hat sich schnell auf die Deutsche Post als Lead-User konzentriert; der Kunde ist anspruchsvoll und bringt die Größe mit, um einen eigenen Markt zu etablieren; gemeinsam wurden sehr individuelle Lösung entwickelt.

► Erst mit der Übernahme durch die Deutsche Post konnte der StreetScooter in den Vertrieb an Dritte gehen; dank der Deutschen Post sogar mit praktischen Erfahrungen im Flotteneinsatz; StreetScooter ist jetzt Experte für Elektrofahrzeuge und Flotten.

► Der Einsatz des StreetScooters wurde sehr intensiv betrachtet; dahinter standen umfangreiche Analysen, Befragungen, viel persönlicher Kontakt, ebenso Praxisbezug, da die Zusteller häufig im Alltag begleitet wurden.

► Das Feedback der Zusteller wurde nicht nur aufgenommen, sondern es wurden davon reale Verbesserungen abgeleitet.

Risk management – valuable insights from the development of StreetScooter

► StreetScooter quickly focused on the Deutsche Post as lead user; the customer is demanding and has the size to establish its own market; a highly customized solution was collaboratively developed.

► Only upon the acquisition by the Deutsche Post was the StreetScooter able to enter into sales to third parties; thanks to the Deutsche Post with its practical experience in fleet operations, StreetScooter is now an expert on electric vehicles and fleets.

► The use of StreetScooter was considered very intensively; behind it stood extensive analyses, surveys, lots of personal contact, as well as practical experience, as the deliverers were often accompanied in their everyday work.

► The feedback from the deliverers was not only recorded, but real improvements were derived from it.

Grundsätzliches Fazit:

Das technische Risiko muss kontinuierlich bewertet werden. Sind frühe Primotypen im Einsatz, kommt dieser Aufgabe eine besondere Bedeutung zu. Daneben ist jedoch das Marktrisiko von genauso hoher Bedeutung. Auch das Marktrisiko muss früh bewertet und minimiert werden. Dazu bieten sich sogenannte Expeditionsmärkte an. Im ersten Schritt muss für die jeweilige Technologie ein Lead-User identifiziert werden. Für diese muss ein kontextspezifisches Expeditions-Design entwickelt werden. Also ein Produkt, das auf die Lead-User zugeschnitten ist. Die Anwendung des Produkts durch die Lead-User muss schließlich genau begleitet und ausgewertet werden. So kann das Produkt aus einer Marktnische heraus entwickelt werden. Ganz wie beim StreetScooter und der Deutschen Post.

Basic conclusion:

The technical risk must be continuously assessed. If early primotypes are used, this task is of particular importance. In addition, however, the market risk is of an equally great importance. The market risk must also be assessed and mitigated early on. Expedition markets are particularly suitable. In the first step, a lead user must be identified for each technology. A context-specific expedition design needs to be developed for this. In other words, a product that is tailored to the lead user. The application of the product by the lead user must then be closely monitored and evaluated. Thus, the product can be developed from a market niche outwards. Just like with StreetScooter and the Deutsche Post.

DISRUPTIVE NETWORK APPROACH (DNA)

BUILDING BLOCK 3

DISRUPTIVE NETWORK APPROACH (DNA)

Im vorigen Kapitel (Lösungsbaustein 2 „Customer Engineering") wurde die Bedeutung der frühzeitigen Kundenintegration in den Entwicklungsprozess beleuchtet. Dabei ging es vor allem um das Thema der effizienten Individualisierung eines Produktes, aufgezeigt am Beispiel des Musterfalles StreetScooter.

Der „Disruptive Network Approach" als dritter Baustein im Rahmen der RoE-Methodik basiert auf der Erkenntnis, dass sich die eingefahrenen Kommunikationspfade, wie sie in klassischen Zulieferstrukturen begangen werden, wenig innovationsfördernd auswirken. Die StreetScooter-Innovation ging und geht andere Wege und setzt auf Leistungserstellung und Informationsbeschaffung im Netzwerk. Die Kernaussage lautet: Produktentwicklung ist als Konfigurationsprozess im Netzwerk mit fertigen Lösungen aus dem Netzwerk zu sehen. Der Lösungsbaustein 3 besteht aus 3 Lösungsprinzipien: Duale Netzwerkstrategie, Side Loading, Kooperationsmanagement.

In the preceding chapter (Building Block 2 "Customer Engineering") the importance of early customer integration in the development process was highlighted. This chiefly concerned the issue of efficient customization of a product shown in the example of the model case of StreetScooter.

The Disruptive Network Approach as a third building block within the RoE methodology is based on the realization that the tried and true communication paths, such as those taken in classic supply structures, do little to foster innovation. The StreetScooter innovation took and takes other paths and focuses on provision of service and information gathering in the network. The key message is: Product development is seen as a configuration process in the network with ready-made solutions from the network. Building block 3 consists of 3 principles: dual network strategy, side loading, and cooperation management.

LÖSUNGSPRINZIP 3.1.

Duale Netzwerkstrategie

Der Disruptive Network Approach[32] (kurz DNA) stellt den Kunden in den Fokus aller Aktivitäten; alles ist auf den Kundenwert ausgerichtet. Um den Kundenwert optimal zu treffen, gilt es, genau die richtigen Partner für die Leistungserstellung im Netzwerk an Bord zu holen.

PRINCIPLE 3.1.

Dual network strategy

The Disruptive Network Approach[32] (DNA) places the customer at the center of all activities; everything is focused on customer benefit. In order to optimally achieve the customer value, it is important to accurately onboard the right partner for the provision of services on the network.

32 Disruptive Network Approach, Methode zur Entwicklung Disruptiver Innovationen; dabei handelt es sich um Innovationen, die eine bestehende Technologie, ein etabliertes Produkt oder eine bekannte Dienstleistung möglicherweise vollständig verdrängen
Disruptive Network Approach: method for the development of disruptive innovations; these are innovations that may completely supplant an existing technology, an established product or service

Zwei Strategien in einem: die duale Netzwerkstrategie

Nach der Theorie von Michael E. Porter[33] sind zwei Strategien zur Erlangung der Marktführerschaft erfolgversprechend: Kostenführerschaft oder Differenzierungsstrategie. Bei der dualen Netzwerkstrategie geht es jedoch um die Kombination beider Wege. Diese Strategie lässt sich anschaulich anhand des Beispiels der Singapore Airlines erläutern. Die Doppelstrategie der Fluggesellschaft setzt sich zusammen aus Qualitätsführerschaft plus Kostenführerschaft. Singapore Airlines genießt einen exzellenten Ruf in der Luftfahrtbranche und steht für qualitativ hochwertigen Kundenservice trotz eines außerordentlichen Konkurrenzdrucks. Der Erfolg, der aus dieser Doppelstrategie resultiert, zeigt, dass Kostenfokus und Qualitätssicherung nicht im Widerspruch stehen müssen.

Zurück zur Auswahl der Netzwerkpartner. In Abhängigkeit des Kundennutzens der Komponenten sind die jeweiligen Zulieferer auszuwählen. Für Komponenten mit einem hohen Kundennutzen sind Qualitätsführer auszuwählen und für Komponenten mit einem niedrigen Kundennutzen sind es die Kostenführer, die zu selektieren sind. Im Fokus muss dabei immer das Kosten-Nutzen-Verhältnis aus Sicht des Kunden stehen. Gerade für deutsche Unternehmen mit hohem Qualitätsanspruch stellt dies eine Herausforderung dar. In einer Umfrage[34] der RWTH zeigte sich, dass zwar ersteres funktioniert, ein systematisches Downsizing aber nicht stattfindet. Entscheidend ist dabei das Downsizing

Two strategies in one: the dual network strategy

According to the theory of Michael E. Porter[33], there are two promising strategies for achieving market leadership: Cost leadership or differentiation strategy. The dual network strategy, however, involves the combination of the two paths. The case of Singapore Airlines provides a clear illustration of this strategy. The airline's dual strategy consists of quality leadership plus cost leadership. Singapore Airlines has an excellent reputation in the aviation industry and stands for high quality customer service in spite of extraordinary competitive pressure. The success resulting from this dual strategy shows that cost focus and quality assurance need not contradict each other.

Back to the selection of network partners. Suppliers are to be selected based on the customer benefit of the respective components. For components with a high customer benefit, quality leaders should be selected and components with a low customer benefit are the cost leaders, who are to be selected. The focus must always be the cost/benefit ratio from the customer's perspective. This represents a particular challenge for German companies with high quality standards. In a survey[34] the RWTH showed that while the former works, a systematic downsizing does not take place. The decisive factor is the downsizing of components, thus reducing costs for components with low customer benefit. This principle applied to the example StreetScooter means: investing a lot in a

33 Michael E. Porter, US-amerikanischer Ökonom und Universitätsprofessor für Wirtschaftswissenschaft am Institute for Strategy and Competitiveness an der Harvard Business School. Er gilt weltweit als einer der führenden Managementtheoretiker Michael E. Porter, American economist and professor of economics at the Institute for Strategy and Competitiveness at Harvard Business School. He is known worldwide as one of the leading management theorists

34 Quelle: https://www.apprimus-verlag.de/herausforderungen-disruptiver-innovationen-am-beispiel-der-elektromobilitat.html
Source: https://www.apprimus-verlag.de/herausforderungen-disruptiver-innovationen-am-beispiel-der-elektromobilitat.html

von Komponenten, also Kostenreduktion für Komponenten mit niedrigem Kundennutzen. Dieses Prinzip am Beispiel StreetScooter bedeutet: viel investieren für ein qualitativ hochwertiges Kamerasystem zum Rangieren; wenig investieren für einen qualitativ ausreichenden und zuverlässigen Elektromotor. Während der Ansatz prinzipiell einfach ist, ist dessen Umsetzung deutlich schwieriger. Schon die richtige Auswahl der Netzwerkpartner findet meist nicht unter strategischen Gesichtspunkten statt. Dies belegt auch eine Studie[35] der RWTH: 85 % der Befragten messen dem Netzwerk eine sehr hohe Bedeutung bei der Entwicklung von Elektromobilität bei. Aber nur 17 % der Befragten wählen die Netzwerkpartner nach strategischen Prämissen aus.

Auswahl der Netzwerkpartner – wertvolle Erkenntnisse aus der StreetScooter-Entwicklung

▶ Komplementarität: Es ist zu prüfen, ob die strategischen Erfolgspositionen und -potenziale der Netzwerkpartner zu einander passen, sich ergänzen und die Partnerschaft so einen Mehrwert erzeugt.

▶ Kompatibilität: Es ist zu prüfen, ob Struktur, Kultur und Motive der Netzwerkpartner zu einander passen und möglichst deckungsgleich sind; unternehmenskulturelle Differenzen sind zu vermeiden.

high-quality camera system for maneuvering and investing little in a reliable and qualitatively sufficient electric engine. While the approach is in principle simple, its implementation is considerably more difficult. Even the correct selection of network partners rarely takes place from a strategic standpoint. This is also proven in an RWTH study[35]: 85 % of the respondents attribute a very high importance to the network for the development of electric mobility. But only 17 % of the respondents choose the network partners based on strategic premises.

Selection of the network partners – valuable insights from the development of StreetScooter

▶ Complementarity: it is necessary to consider whether the strategic positions and potentials of the network partners match each other, complement each other and that the partnership creates an added value.

▶ Compatibility: it is necessary to consider whether structural, cultural and motives of the network partners are compatible with each other and are as congruent as possible; corporate cultural differences are to be avoided.

35 Quelle: https://www.apprimus-verlag.de/herausforderungen-disruptiver-innovationen-am-beispiel-der-elektromobilitat.html
Source: https://www.apprimus-verlag.de/herausforderungen-disruptiver-innovationen-am-beispiel-der-elektromobilitat.html

Grundsätzliches Fazit:

Im Rahmen der StreetScooter-Entwicklung erfolgt die Auswahl der Netzwerkpartner (Premium-Partner sowie Kostenführer) dual: Die Auswahl der Premium-Partner erfolgt mit Blick auf die Merkmale mit direktem Kundennutzen (Beispiel: hochauflösende Kamerasysteme zum Rangieren); die Auswahl der Kostenführer erfolgt hinsichtlich anderer Merkmale (Beispiel: Space-Frame-Struktur[36] der Karosserie).

Wichtig in Bezug auf die Zusammenarbeit ist generell die kontinuierliche Weiterentwicklung des Netzwerks zur Schärfung des Profils.

LÖSUNGSPRINZIP 3.2.

Side Loading

Erfolgt der Entwicklungsprozess auf der Basis des Disruptiven Approaches sowie der Konfigurationsprozess mit fertigen Modulen, wird sehr bald deutlich, dass es eine stärkere Trennung zwischen Modul- und Produktentwicklung geben muss.

Das Prinzip des Side Loadings im Fahrzeugbau

Fahrzeuge werden heute über ihren Lebenszyklus grundsätzlich nicht verändert. Nach einer

Basic conclusion:

As part of the StreetScooter development, the selection of network partners (premium partners and cost leaders) was twofold: The selection of premium partners was done with a view to the characteristics with direct customer benefit (for example, high-resolution camera systems for maneuvering); the selection of the cost leaders was done based on other characteristics (for example, space frame structure[36] of the body).

In terms of the collaboration, the continuous development of the network for raising the profile is generally critical.

PRINCIPLE 3.2.

Side loading

If the development process is based on the disruptive approaches and the configuration process with finished modules, it will quickly become obvious that there must be a clearer separation between module and product development.

The principle of side loading in vehicle manufacturing

Today, vehicles are essentially not changed during their lifecycle. After a certain time, "facelifts" are

36 Space-Frame, ist eine Karosseriebauart in Skelett-Struktur; die Karosserie besteht aus geschlossenen Hohlprofilen, die direkt oder über Knoten miteinander verbunden sind. Flächige Bauteile (wie Dach oder Windschutzscheibe) werden zur Aufnahme von Schubkräften steif mit dem Skelett verbunden. So können neue Materialien, Teile und Techniken im Karosseriebau verwendet werden
Space frame is a body type in skeletal structure. The body is made of closed hollow profiles, which are connected directly or via nodes. Flat components (such as roof or windshield) are rigidly connected to the skeleton in order to absorb thrust. This way, new materials, components and techniques can be used in the car body construction

bestimmten Zeit werden partiell sogenannte Face-Lifts durchgeführt, wobei einzelne Module ausgetauscht werden. Side Loading funktioniert anders. Diesem Prinzip folgend werden Innovationen kontinuierlich in das Fahrzeug eingebracht; aktualisierte Module fließen in die Serie.

Side Loading im Detail

Die Produkt- und die Prozessentwicklung müssen mit der allgemein steigenden Komplexität des Produktprogramms umgehen. Insgesamt steigt die Anzahl der Fahrzeugderivate bei gleichzeitig sinkender Lebensdauer der Modell-Generationen. Durch die Elektromobilität steigt die bereits bestehende Komplexität des Produktprogramms noch deutlich an, unabhängig vom gewählten Entwicklungsansatz (Conversion Design[37] oder Purpose Design[38]). Vor dem Hintergrund steigender Produktkomplexität nimmt auch die Bedeutung der integrierten Betrachtung von Produkten und Prozessen zu. Hier gilt es, Lösungen zu finden, um die Produktkomplexität möglichst effizient in der Produktion abbilden zu können. Je größer die Freiheitsgrade in der Produktentwicklung sind, desto schwerer fällt es der Produktionsplanung, die beste Lösung zur produktionstechnischen Umsetzung ermitteln zu können. Um diesem Problem zu entgehen, ist die Einführung einer übergeordneten Entwicklungsinstanz notwendig: das Side Loading. Beim Side Loading handelt es sich um eine Entwicklungsinstanz, die die zukünftigen

to some extent performed where individual modules are replaced. Side loading, however, works differently. Following this principle, innovations are continuously introduced into the vehicle; updated modules flow into the series.

Side loading in detail

The product and process development needs to deal with the generally increasing complexity of the product portfolio. Overall, the number of vehicle derivates increases with falling service life of model generations. With electromobility, the existing complexity of the product range is significantly rising even more, regardless of the chosen development approach (conversion design[37] or purpose design[38]). Against the background of increasing product complexity, the importance of integrated analysis of products and processes also rises accordingly. Here it is important to find solutions in order to reflect the complexity of products as efficiently as possible in the production. The greater the degree of freedom in the product development, the harder it is for production planning to determine the best solution for technical production implementation. To avoid this problem, the introduction of a superior development entity is necessary: side loading. Side loading is a development entity that adjusts vehicle structures and the structures of the vehicle production, planning for the future, and does so independently of development projects and product lifecycles. This

37 Conversion Design: Entwicklungsansatz bei dem ein neues Elektrofahrzeug auf Basis eines bestehenden Fahrzeugs mit Verbrennungsmotor entwickelt wird. Der Antrieb wird in dem Fall einfach von einem Verbrennungsmotor auf einen Elektroantrieb getauscht.
Conversion Design: Development approach in which a new electric vehicle is developed based on an existing vehicle with an internal combustion engine. In this case, the drive is simply switched from an internal combustion engine to an electric drive.

38 Purpose Design: Entwicklungsansatz bei dem ein Elektrofahrzeug von Grund auf neu und als reines Elektrofahrzeug entwickelt wird. Eine Variante des Fahrzeugs mit Verbrennungsmotor ist in dem Fall nicht vorgesehen.
Purpose Design: Development approach in which an electric vehicle is developed from the ground up and as a pure electric vehicle. In this case, there is no variant of the vehicle with an internal combustion engine.

Fahrzeugstrukturen sowie die Strukturen der Fahrzeugproduktion abstimmt und vordenkt, und zwar unabhängig von Entwicklungsprojekten und Produktlebenszyklen. Hierzu zählen die Entwicklung der Produktarchitektur, der Prozess- und Anlagenbaukästen sowie das Technologiemanagement. Die Entwicklungsergebnisse des Side Loadings werden den Konstrukteuren und Planern wie in einem Supermarkt angeboten, aus dessen Regalen sie sich für eine konkrete Entwicklungsaufgabe bedienen können bzw. müssen.

Die Problematik zu langer Produktlebenszyklen

Die klassischen Produktlebenszyklen betragen in der Automobilindustrie sieben Jahre. Bei einem Entwicklungsvorlauf von drei Jahren sind die in diesem Zeitraum verbauten Technologien am Ende des Lebenszyklus zehn Jahre alt. Bei aktuell rasanter Technologieentwicklung ist ein solcher Zyklus viel zu lang und führt dazu, dass Fahrzeugtechnologien oft sehr schnell veraltet sind.

Anders sieht es dagegen in der IT- und der Mobiltelefoniewelt aus: Hier beträgt der Lebenszyklus eines Produktes im Schnitt nur ein Jahr. Diese Tatsache stellt eine echte Herausforderung für die Fahrzeugentwicklung dar. Das Problem: Vor zehn Jahren gab es noch kein Smartphone. Heute zählt man in Deutschland rund 50 Mio. Smartphone-Nutzer! Entsprechend ist die Konnektivität[39] an Bord eines Autos häufig eine Herausforderung, da sich die Technologie von Mobiltelefonen

includes the development of the product architecture, process and plant construction kits and technology management. The development results of the side loading are offered to designers and planners as in a supermarket, where they can or need to select them from the shelves for a specific development task.

The problem with long product lifecycles

The classic product lifecycle in the automotive industry is seven years. With a development lead-time of three years, the built-in technologies in that period are ten years old at the end of the lifecycle. With the pace of today's rapid technology development, such a cycle is much too long and leads to vehicle technologies often becoming obsolete very quickly.

On the other hand, this situation is different in the IT and mobile telephony world, where the lifecycle of a product is on average just one year. This fact represents a real challenge for vehicle development. The problem: Ten years ago, there was no such thing as a smartphone. Today, there are about 50 million smartphone users in Germany! Accordingly, connectivity[39] on board a car is often a challenge, since cell phone technology has evolved significantly compared to built-in car connectivity solutions. The problem will become even greater with increasing development rates.

39 Konnektivität, steht u. a. für die Schnittstellenausstattung von informationstechnischen Geräten, siehe Hardwareschnittstelle Connectivity in this context refers to the interface equipment of information technology equipment, see Hardware interface

im Vergleich zu der in Autos verbauten Konnektivitätslösungen deutlich weiterentwickelt hat. Das Problem wird mit zunehmenden Entwicklungsgeschwindigkeiten sogar noch größer werden.

Produktentwicklung und Produktentwicklungszyklen – wertvolle Erkenntnisse aus der StreetScooter-Entwicklung

► Modul- und Produktentwicklung müssen voneinander getrennt werden, um das kontinuierliche Updaten von Autos zu ermöglichen.

► Nicht nur einmalige Facelifts im Produktlebenszyklus, sondern regelmäßige Integration neuer Technologien und weiterentwickelter Module in die Serie.

► Anlaufmanagement als Fähigkeit zur Veränderung der Produktion und Einführung neuer Module wird Kernkompetenz.

Grundsätzliches Fazit:

Heute steht ein aufwendiger Anlaufprozess am Beginn eines Produktlebenszyklus, danach wird eine stabile Produktion gefahren, fallweise mit partiellen Eingriffen (Facelifts) über den Lebenszyklus.

Zukünftig werden kontinuierlich Anläufe gefahren, dabei werden ständig Updates in das Fahrzeug einfließen und Innovationen in den Modulen

Product development and product development cycles – valuable insights from the development of StreetScooter

► Module and product development must be separated in order to make the continuous updating of cars possible.

► Not only one-off facelifts in the product lifecycle, but regular integration of new technologies and advanced modules into the series.

► Start-up management as ability to change the production and introduction of new modules becomes core competence.

Basic conclusion:

Today, there is a complex ramp-up process at the beginning of a product lifecycle, after which a stable production is run, in some cases with partial interventions (facelifts) over the lifecycle.

In the future, ramp-ups will be continually run, whereby updates will be constantly integrated into the vehicle and innovations implemented in the modules. Production will change: it will then be in a quasi-permanent ramp-up process. Ramp-up management becomes core competence. It is important to establish this ramp-up competence not only within the company, but also across the entire supply chain, as logistics processes are also always affected and the service likewise needs to

umgesetzt. Die Produktion wird sich wandeln, sie wird sich quasi in einem ständigen Anlaufprozess befinden. Anlaufmanagement wird Kernkompetenz. Wichtig dabei ist, diese Anlaufkompetenz nicht nur im eigenen Unternehmen zu etablieren, sondern über die ganze Lieferkette, da immer auch Logistikprozesse betroffen sind und auch der Service abgedeckt werden muss. Vor dem Hintergrund kontinuierlicher Updates ist zu beachten, dass sich die jeweiligen Veränderungen auf die Straßenzulassung auswirken können und eine neue Homologation[40] erfordern. Dieser erhöhte Aufwand ist nur bei einem entsprechenden Kundenwert gerechtfertigt.

LÖSUNGSPRINZIP 3.3.

Kooperationsmanagement

Produkt-, sprich Fahrzeugentwicklung gemäß DNA, heißt Arbeit im Netzwerk; diese Arbeit bedeutet für alle Partner ein kontinuierliches Zusammenspiel aus ständigem Austausch, permanentem Abstimmen und Verhandeln. Hier ist ein reibungsloses Schnittstellenmanagement gefragt. In der intensiven Zusammenarbeit mit Partnern im Netzwerk kommt daher dem Kooperationsmanagement eine besondere Bedeutung zu. Die Partner müssen dabei grundsätzlich wissen, dass es sich bei dieser Form der Zusammenarbeit um eine Kooperation handelt und nicht nur um ein vertraglich fixiertes Lieferantenverhältnis.

be covered. Against the background of continuous updates, it should be noted that the respective changes to the street legal approval can have an impact and require a new vehicle homologation[40]. This increased expense is justified only with a corresponding customer value.

PRINCIPLE 3.3.

Cooperation management

According to DNA, product development – here, vehicle development – means working in the network. For all partners, this work is a continuous interaction of constant exchange, permanent agreement making and negotiating. This all demands a smooth interface management. In the intensive collaboration with partners in the network, cooperation management is of particular importance. The partners have to fundamentally know that this kind of collaboration is about a cooperation, and not just about a contractually fixed supplier relationship.

40 Homologation, der Begriff besagt, dass Fahrzeuge so konfiguriert werden müssen, dass sie die Straßen-Zulassungsfähigkeit mitbringen. Als Grundlage der Homologation gilt für die Automobilindustrie die nationale Straßenverkehrszulassungsordnung [StVZO]
Homologation: this term means that vehicles must be configured so that they are able to meet street legal authorization standards. For the automotive industry, the German Road Vehicle Registration Regulation (Straßenverkehrszulassungsordnung/StVZO) serves as a basis for homologation.

Die Bedeutung von Kooperationen

„Innovationen entstehen grundsätzlich an Schnittstellen und in der Kombination verschiedener Lösungen, nicht durch eine einzelne fokussierte Fähigkeit"[41]. Es gibt viele Beispiele für innovative Produkte, deren Entwicklung im Wesentlichen auf Netzwerklösungen zurückzuführen ist. Durch die Kooperation der Firma InvenSense, die Bewegungssensoren herstellt mit der Nintendo (Wii), wurden ganz neue Spiele und Anwendungsarten möglich. Der Kernsatz lautet: Um distruptive Innovationen zu lancieren, ist ein Zusammenspiel sich ergänzender Partner unerlässlich. Vor diesem Hintergrund – und mit Blick auf eine zunehmende Spezialisierung in der Technologieentwicklung – kommt mehr und mehr ein allgemeiner Trend zur Zusammenarbeit in Netzwerken in Gang. Zahlreiche Studien belegen diese Entwicklung. Dabei lassen sich statistische Signifikanzen zwischen der Intensität der Arbeit in Netzwerken und dem Innovationsgrad der Beteiligten aufzeigen. Je stärker Unternehmen kooperieren, desto innovativer sind sie. Die Relevanz von Kooperationen steigt vor allem bei neuen Technologien. Denn hier müssen die fehlenden Kompetenzen eines einzelnen Unternehmens besonders häufig durch Netzwerkpartner ausgeglichen werden. Auch lassen sich Entwicklungen so schneller vorantreiben. Letztendlich entwickeln gut geführte Netzwerke nicht nur Innovationen, sondern auch „Antennen" für neue Ideen und Lösungen.

The importance of cooperations

"Innovations essentially occur at interfaces and in the combination of different solutions, not by a single focused ability"[41]. There are many examples of innovative products whose development can essentially be traced back to network solutions. The cooperation of the company InvenSense made it possible for the motion sensors manufactured with the Nintendo (Wii) to become entirely new games and application types. The key here is this: To launch disruptive innovation, having an interplay of complementary partners is essential. Against this background – and with a view to an increasing specialization in technology development – there is ever-growing overall trend towards collaboration in networks. Numerous studies confirm this development. These highlight statistic significances between the intensity of working in networks and the degree of innovation of the parties. The more companies cooperate, the more innovative they are. The relevance of cooperation increases especially with new technologies. With these, the competencies one individual company may lack can most often be offset by those of network partners. This also allows developments to advance that much faster. Ultimately, well-managed networks not only develop innovation but they also develop "antennas" for new ideas and solutions.

41 Zitat: „Innovation occurs at the boundaries between mind sets, not within the provincial territory of one knowledge and skill base", Leonard-Barton-Group, 1995
Quote: "Innovation occurs at the boundaries between mindsets, not within the provincial territory of one knowledge and skill base," Leonard-Barton Group, 1995

Innovation durch Kooperation – wertvolle Erkenntnisse aus der StreetScooter-Entwicklung

► Dezentralität: Netzwerkpartner sind gleichberechtigt; letztendlich zählt der Kundenwert; Spezifikationen sind im Sinne des Gesamtoptimums zwischen den Partnern verhandelbar; Produktentwicklung und Produktion arbeiten integriert.

► Heterogenität: Unterschiedliche Unternehmen bereichern ein Netzwerk mit unterschiedlichen Fähigkeiten; große etablierte Unternehmen bringen viel Erfahrung mit, kleine mittelständische Unternehmen haben einen effizienten Pragmatismus und schnelle Entscheidungen; unterschiedliche Partner können sich in einer Kooperation gegenseitig befruchten.

► Offenheit: offen sein für neue Ideen, neue Partner, neue Ansätze, Methoden, neue Technologien; Bestehendes hinterfragen; den Willen zur Veränderung mitbringen.

Grundsätzliches Fazit:

Das Unternehmen StreetScooter ist als großes, offenes und flexibles Netzwerk gestartet. Von Anfang an wurde ein intensives Kooperationsmanagement betrieben. Dezentralität, Heterogenität und Offenheit waren entscheidend für den Erfolg. Das Thema Dezentralität war für die Abstimmung

Innovation through cooperation – valuable insights from the development of StreetScooter

► Decentralization: Network partners have equal rights; ultimately what counts is customer value; specifications are negotiable in terms of the overall optimum between partners; product development and production work hand in hand.

► Heterogeneity: Different companies enrich a network with different abilities; large established companies bring a lot of experience, small and middle-sized enterprises (SMEs) have an efficient pragmatism and can make quick decisions; in a cooperation, different partners can cross-fertilize.

► Openness: being open to new ideas, new partners, new approaches, methods, new technologies; questioning the status quo; bringing the will to change.

Basic conclusion:

The company StreetScooter started as a large, open and flexible network. From the beginning, there was an intensive cooperation management in place. Decentralization, diversity and openness were essential for its success. The issue of decentralization was especially important for the coordination and optimization of the design and for the

und Optimierung des Designs mit der Konstruktion sowie für den Werkzeugbau besonders wichtig. In vielen Schritten passten die Partner das Produkt so lange an, bis das Design passte, die Konstruktion stimmte und die Werkzeugkosten minimal waren. Ein klarer Beleg für Heterogenität ist die Tatsache, dass im Netzwerk alles verbunden war und miteinander gearbeitet hat, vom KMU aus der Region bis zum größten Automobilzulieferer. In vielen Workshops entwickelten sich spannende Diskussionen, da oft sehr unterschiedliche Welten aufeinandertrafen, deren Vertreter aber voneinander lernen wollten und konnten. Bewusst wurden dabei auch die Anlagenbauer in diese Diskussionen einbezogen, da sie oft über tiefe Einblicke in Produktionsprozesse verfügen. Und letztendlich haben sich die Kooperationen im Umfeld immer sehr dynamisch entwickelt. Dank einer grundsätzlichen Offenheit kamen über die Zeit viele neue Partner hinzu.

toolmaking. The partners continually modified the product in many steps, until the design was right, the construction was aligned, and the tooling costs were minimal. A clear evidence of heterogeneity is the fact that everyone was connected and worked together in the network, from the SMEs from the region to the largest automotive suppliers. There were many charged discussions held in many workshops, where very different worlds often clashed, but those from these worlds nonetheless shared the desire to learn from each other and did so. A conscious effort was made to also involve plant engineers in these discussions, since they often have deep insight into production processes. And finally, the collaborations in the field always developed extremely dynamically. In time, many new partners were added thanks to this fundamental openness.

DIE PERSPEKTIVE
THE PERSPECTIVE

BEGRIFFE WIE KOMPLEX UND KOMPLIZIERT WOLLEN UNS DIE GRENZEN DER VORGEHENSWEISEN

IN SYSTEMISCHEN PROZESSEN SIGNALISIEREN. OFT SIND SIE ABER NUR AUSREDEN.

Prof. Achim Kampker
Geschäftsführer der Streetscooter GmbH
Executive Vice President E-Mobilität Deutsche Post DHL
CEO of Streetscooter GmbH
Executive Vice President E-Mobility Deutsche Post DHL

7

DIE STREETSCOOTER-NETZWERK-STORY

THE STREETSCOOTER NETWORK STORY

Innovationen entstehen vor allem dann, wenn man über den eigenen Tellerrand hinausschaut, vor allem aber, wenn man Dritten über diesen Tellerrand den Einblick nach innen erlaubt und Kunden integriert. Genauso haben es die Ingenieure des StreetScooters gemacht. Hier ein Einblick und Überblick.

Innovations mostly come about when you look outside your box, but especially when you allow third parties outside that box to have an insight into it and when you integrate clients. That's just what the engineers of the StreetScooter did. Here is an insight and an overview.

MIT DEM STROM, ABER GEGEN ALTE PRINZIPIEN

WITH THE FLOW, BUT AGAINST OLD PRINCIPLES

Wie der StreetScooter zur idealen „Beziehungskiste" für Partner-Integration und effiziente Zusammenarbeit wurde? Ganz einfach. Man kann eine Idee im Geheimen entwickeln und eines Tages aus der Schublade ziehen. Auf diesem Weg muss man Entwicklungszeit und Entwicklungskosten selber stemmen. Oder man kann sofort mit Partnern an einer Idee arbeiten, um auf effizientere Art und Weise einen früheren Markteintritt zu schaffen. Der Vorteil dieser Strategie liegt auf der Hand: Wer schneller im Markt ist, kann auch schneller Geld verdienen.

Netzwert-Partnerschaft statt Netzwerk-Partnerschaft

Am Anfang war StreetScooter für das Netzwerk vor allem eines: eine Technologieplattform, auf der die Netzwerkpartner ihre Technologien demonstrieren und testen konnten. So war StreetScooter für alle Netzwerkpartner immer mehr als ein Projekt. Durch die Zusammenarbeit konnten die einzelnen Komponenten der Netzwerkpartner im Gesamtsystem getestet werden, den Kunden konnte mit StreetScooter schnell ein erstes Referenzfahrzeug präsentiert werden und die Offenheit für neue Lösungen ermöglichte zahlreiche Innovationen. Überhaupt war die Transparenz von Spezifikationen und technischen Schnittstellen für viele Partner Motivation, sich an StreetScooter zu beteiligen. Schnittstelleninformationen, die normalerweise im Fahrzeug verschlossen sind, wurden offengelegt.

How did the StreetScooter become the perfect "relationship" for partner integration and efficient cooperation? Very easily. You can secretly develop an idea and then one day pull it out of your drawer. In this way, you have to cover development time and development costs yourself. Or, you can immediately start working with partners to more efficiently get it to the market sooner. The advantage of this strategy is obvious: He who gets to market faster, can also earn money faster.

Network value partnership instead of network partnership

Initially, StreetScooter was pretty much just one thing for the network: a technology platform on which network partners could demonstrate and test their technologies. So, for all network partners, StreetScooter was always more than a project. Through collaboration, the individual components of the network partners could be tested in the overall system. With StreetScooter, the customer could quickly be presented with a first reference vehicle and the openness to new solutions led the way to numerous innovations. In general, the transparency of specifications and technical interfaces was a motivator for many partners to participate in StreetScooter. Interface information that is normally closed off in the vehicle was disclosed.

This aspect of the cooperation alone already created value for network partners in terms of their

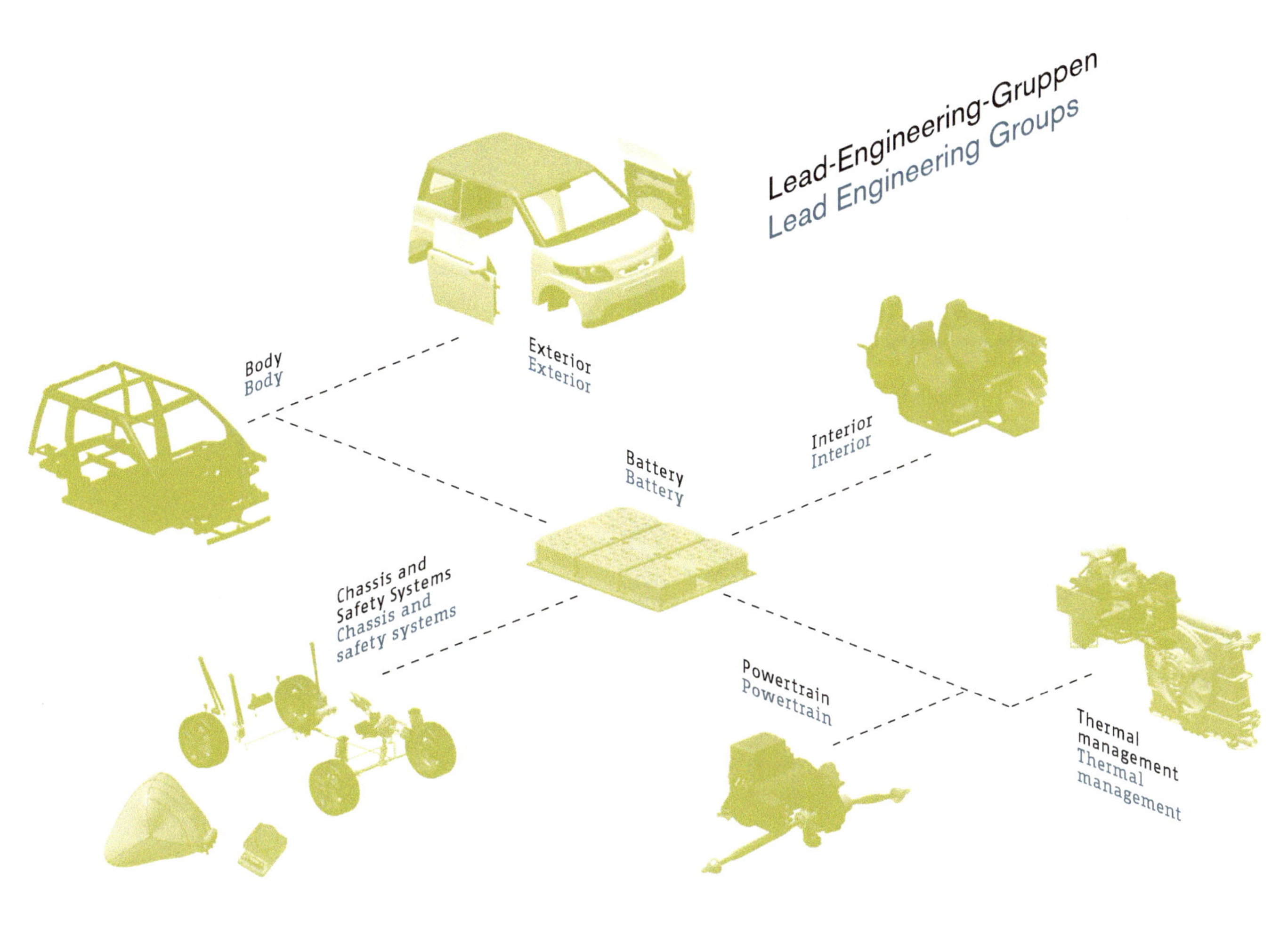

Lead-Engineering-Gruppen
Lead Engineering Groups
Body
Body
Exterior
Exterior
Interior
Interior
Battery
Battery
Chassis and
Safety Systems
Chassis and
safety systems
Powertrain
Powertrain
Thermal
management
Thermal
management

So entstand schon aus der Zusammenarbeit ein Wert in Bezug auf Innovationsfähigkeit der Netzwerkpartner. Aus der Netzwerk-Partnerschaft wurde eine Netzwert-Partnerschaft.

Innovativ durch schnellen Austausch im Netzwerk

80 Partner in ein neues Netzwerk integrieren erforderte vor allem eins: Kommunikation. Jeder Partner hatte zu Beginn eine Erwartungshaltung und eine Vorstellung über die Zusammenarbeit im Netzwerk. Diese unterschiedlichen Perspektiven zu integrieren war die erste Herausforderung. Anforderungen aufnehmen, Schnittstellen identifizieren, Ziele priorisieren – alles dies war erst mal nötig, um alle Partner abzuholen. Besonders herausfordernd war die Tatsache, dass dazu noch mit offenen Spezifikationen gearbeitet wurde. Nicht alle Rahmenbedingungen waren gesetzt. Viele Schnittstellen zwischen Partnern mussten erst definiert werden.

Nach dem ersten Zusammenfinden musste eine gemeinsame Sprache gefunden werden. Jedes Unternehmen hatte seinen eigenen Erfahrungshintergrund und benutzte teilweise ganz eigene Begriffe. Vom großen Automobilzulieferer bis zum kleinen Unternehmen aus der Region waren ganz unterschiedliche Partner im Netzwerk vertreten. Hier eine gemeinsame Kommunikationsbasis zu schaffen war enorm wichtig. Und die Lösung dazu lag nicht in den Kommunikationsmitteln, sondern stand in der Garage. Der erste Prototyp schaffte,

innovation capacity. From the network partnership, a network value partnership was born.

Innovative thanks to fast in-network sharing

80 partners integrated into a new network required one thing above all: communication. At the start, each partner had its own expectation and idea about the cooperation within the network. Integrating these different perspectives was the first challenge. Incorporating requirements, identifying interfaces, prioritizing goals – all this was first needed to pick up all the partners at their starting points. Particularly challenging was the fact that they were still working with open specifications. Not all scope conditions were set. Many interfaces between partners had to first be defined.

After the first gathering, a common language had to be found. Each company had its own background of experience and to some extent used terms that were entirely their own. Very different partners were represented in the network, from the large automotive supplier to the small regional company. Providing them all with a common basis for communication was extremely important. And the solution wasn't in the means of communication, but in the garage. The first prototype managed to do what email, etc. couldn't do. One communication with one language. Because the hardware made it clear what was being discussed. It gave

Arndt G. Kirchhoff
Geschäftsführer KIRCHHOFF Automotive GmbH,
Iserlohn
*CEO KIRCHHOFF Automotive GmbH,
Iserlohn, Germany*

„Die Entwicklung des StreetScooters war bisher insofern ein einmaliges Erlebnis, da ausschließlich Zulieferer, ohne einen Automobilhersteller, zusammengearbeitet haben, um ein Benchmark-Projekt für die urbane Elektromobilität zu realisieren. Das war nicht nur zeit- und kostengünstig, sondern hat auch die nötigen Freiheitsgrade gebracht, ohne die sonst üblichen Vorgaben eines Autoherstellers ein neuartiges Fahrzeugkonzept zu entwickeln. Ich bin davon überzeugt, dass wir in Kürze den Durchbruch der Elektromobilität erleben, da jetzt einerseits die notwendige Infrastruktur gebaut wird und andererseits auch die gesetzlichen Voraussetzungen und Rahmenbedingungen geschaffen werden. Die Deutsche Post ist mit gutem Vorbild vorangegangen, da sie in ihren Betriebshöfen ohne die öffentliche Hand eine Infrastruktur errichten kann. Der Leitmarkt wird in Kürze China sein, da im neuen Fünfjahresplan vorgesehen ist, dass bei Bauvorhaben zukünftig 80 % der auszuweisenden Parkplätze mit Elektroanschluss zu bauen sind und Voraussetzung für die Zulassung eines Fahrzeugs mit Verbrennungsmotor das gleichzeitige Angebot eines batterieelektrischen Autos vorgegeben wird. Die strategische Ausrichtung unserer Automobilhersteller zum konsequenten Ausbau des Angebots von Elektromobilität wird auch in Deutschland zu einem Aufschwung führen. Deutschland ist bisher Leitanbieter mit mehr als 30 Fahrzeugmodellen in der Welt."

"The development of StreetScooter had so far been a unique experience, since suppliers – on their own without a car manufacturer – worked together to realize a benchmark project for urban electric mobility. This not only saved time and money, but it also brought with it the necessary degrees of freedom, freedom from the usual requirements car manufacturers have for developing a new vehicle concept. I am convinced that we're on the threshold of a breakthrough in electric mobility, since on the one hand now the necessary infrastructure is being built, and on the other hand, the legal requirements and conditions are being created. The Deutsche Post has set a good example, since it can establish an infrastructure in its own depots, without the need of the public sector. The lead market will soon be China, since the new Chinese 5-year plan sets forth that in the future, construction projects must build 80 % of reportable parking lots with electrical connections and that a prerequisite for registering a vehicle with a combustion engine is the simultaneous offer of a battery-powered electric car. The strategic orientation of our automakers towards the consistent expansion of the supply of electric mobility will lead to an upswing in Germany too. Germany has so far been the worldwide lead provider, with more than 30 vehicle models."

was E-Mail & Co. nicht schaffen konnten. Eine Kommunikation mit einer Sprache. Denn mit der Hardware war klar, worüber geredet wurde. Es gab was zum Anfassen und Zeigen. Die Schnittstellen wurden sichtbar. Aus dem großen virtuellen Projekt wurden abgegrenzte greifbare Aufgaben.

Je weiter das Projekt voranschritt, desto wichtiger wurde der Change-Prozess. Eine Änderung einer Einzelkomponente kann weiterreichende Folgen im Auto haben. Transparenz über neue Veränderungen und deren Auswirkungen wurde essentiell. Dieser Herausforderung wurde nicht nur systemisch durch das PLM-System[42] begegnet, sondern auch durch einen hohen persönlichen Austausch. Die Netzwerkpartner sind regelmäßig in Workshops auch persönlich zusammengekommen und standen im Austausch.

So schnell kann's gehen: Die Integration der Deutschen Post

Nicht nur die Kommunikation der Netzwerkpartner war entscheidend. Für den Erfolg von StreetScooter noch wichtiger war die Kommunikation mit dem Kunden. Im September 2012 hatte das Bonner Unternehmen einen Entwicklungsauftrag an die StreetScooter GmbH vergeben. Nach nur einjähriger Entwicklungsphase erfüllte das erste Referenz-Modell des StreetScooters die Erwartungen an Ausstattung, Ladekapazität und Sicherheitsstandards der Deutschen Post. 150 DHL-Zusteller hatten an der Entwicklung des „gelben"

42 Product Lifecycle Management (PLM) steht für das Managen aller Daten und Informationen eines Produkts über den gesamten Lebenszyklus. Es existiert eine zentrale Datenquelle [Single Source of Truth], mit der alle anderen Systeme [wie CAD oder CRM] verknüpft sind.
Product Lifecycle Management (PLM) refers to managing all data and information of a product throughout the entire life cycle. There is one central data source (Single Source of Truth), to which all other systems (such as CAD or CRM) are linked.

them all something to touch and show. The interfaces became visible. Out of the large virtual project came defined tangible tasks.

The further along the project progressed, the more important the change process became. A change in a single component can have far-reaching consequences in the car. Transparency about new changes and their impact was essential. This challenge was not only systemically addressed by the PLM system[42], but also by a high level of personal exchange. The network partners also regularly met in workshops in person and exchanged thoughts and views.

It can happen this fast: The integration of Deutsche Post

The communication of the network partners wasn't the only critical factor. For the success of StreetScooter, the communication with the customer was even more important. In September 2012, the Bonn-based company had awarded a development contract to StreetScooter GmbH. After just one year of development, the first reference model of StreetScooter met Deutsche Post's expectations with respect to equipment, loading capacity and safety standards. In 150, DHL couriers had participated in the development of the "yellow" StreetScooter, which would be the Post's first electric delivery vehicle. Their advice from experience ensured, for example, that the designers made the loading surface slip-resistant, that the cargo space had no

StreetScooters, der das erste Elektro-Zustellfahrzeug der Post werden sollte, teilgenommen. Ihre Ratschläge aus der Praxis sorgten zum Beispiel dafür, dass die Konstrukteure die Ladefläche rutschfest machten, dass der Laderaum keine Radkästen hatte und von drei Seiten zugänglich war. Wie schon mit den Netzwerk-Partnern stellten sich die Aachener auch mit allen Beteiligten der Post immer wieder die Frage, wie sich diese oder jene Spezifikation auf die Kosten auswirken würde. Diese Offenheit und Transparenz dem Kunden gegenüber sorgte dafür, dass das Fahrzeug ideal auf die Anforderungen der Deutschen Post zugeschnitten werden konnte. Und so entstand letztendlich auf beiden Seiten ein Mehrwert durch einen zufriedenen Kunden. Also wieder die Netzwert-Partnerschaft.

Der Kooperationsansatz

Nicht die DNA, sondern der DNA. Gemeint ist damit ein radikal anderer Kooperationsansatz, der letztendlich aus der Zusammenarbeit im Netzwerk bei StreetScooter entstanden ist. Der DNA – sprich Disruptive Network Approach[43] – bricht mit vorhandenen Netzwerkstrukturen in der Automobilindustrie, die an einer zentralisierten Pyramidenstruktur orientiert sind. Die wesentlichen Prinzipien wurden bereits in Kapitel 6 vorgestellt: Duale Netzwerkstrategie, Side Loading und Kooperationsmanagement. Die starke Netzwerkorientierung hat bei StreetScooter vor allem zu einem geführt: Schnelligkeit.

wheel arches and that it would be accessible from three sides. As they had already done with the network partners, the Aacheners, together with all stakeholders, constantly asked themselves the question of how this or that specification would affect the cost. This openness and transparency towards the customer made sure that the vehicle could be ideally suited to the requirements of Deutsche Post. And this ultimately created an added value on both sides by having a satisfied customer. So once again, a network value partnership.

The Cooperation Approach

Not the DNA, but the "DNA". This refers to a radically different approach to cooperation that ultimately emerged from the work in network at StreetScooter. The "DNA" – Disruptive Network Approach[43] – breaks with existing network structures in the automotive industry which are based on a centralized pyramid structure. Its key principles were already presented in Chapter 6: Dual network strategy, side loading, and cooperation management. StreetScooter's strong network orientation has led to one thing above all else: speed.

43 Disruptive Network Approach, Methode zur Entwicklung disruptiver Innovationen; dabei handelt es sich um Innovationen, die eine bestehende Technologie, ein etabliertes Produkt oder eine bekannte Dienstleistung möglicherweise vollständig verdrängen
Disruptive Network Approach: method for the development of disruptive innovations; these are innovations that may completely supplant an existing technology, an established product or service

„WIR WOLLTEN DEMONSTRIEREN, DASS ES MÖGLICH IST, IN KURZER ZEIT, MIT EINEM KLEINEN TEAM UND WENIG INVESTITION, EIN FAHRZEUG IN SERIE ZU BRINGEN."

"WE WANTED TO DEMONSTRATE THAT IT IS POSSIBLE TO ACCOMPLISH THE SERIAL PRODUCTION OF A VEHICLE WITHIN A SHORT TIME WITH A SMALL TEAM AND LITTLE INVESTMENT."

Dipl.-Ingenieur Christoph Deutskens
Oberingenieur, Chair of Production Engineering of
E-Mobility Components (PEM) an der RWTH Aachen
und Geschäftsführer PEM Aachen GmbH
*Senior Engineer, Chair of Production Engineering
of E-Mobility Components (PEM) at RWTH Aachen
and Managing Director of PEM Aachen GmbH*

Wo hat die Idee StreetScooter ihre Wurzeln?

Christoph Deutskens: Die Idee entstand bei uns am Institut. Das war 2008, 2009, als das Thema Elektromobilität immer relevanter wurde. Wir haben erörtert, welchen Einfluss diese Entwicklungen auf die Automobilindustrie haben würden. Schon damals war klar, dass die gesamte Automobilindustrie sich sehr stark verändern und folglich mit massiven Umbrüchen konfrontiert sein würde. Zuerst war also das Thema Elektromobilität da. Gleichzeitig entstand die Frage, wie man Forschung und Praxis näher zusammenbringen könnte. Das brachte uns zu der Einsicht, dass wir näher an der Praxis operieren müssten, um selber eine Innovation voranzutreiben. Dafür benötigten wir ein Use-Case, ein Fahrzeug, um daran Dinge demonstrieren zu können. Zu diesem Zeitpunkt war schon klar, dass die E-Fahrzeuge sich unmöglich durchsetzen können, solange sie bei Weitem kostspieliger sind als konventionelle Fahrzeuge. So ist die Idee entstanden: Wir wollen ein bezahlbares Elektroauto entwickeln.

Welches Ziel verfolgten Sie zu Beginn?

Die Vision war, ein Auto für 5.000,– Euro zu entwickeln.

Herstellungskosten?

Die 5.000,– Euro mussten ausreichen, um ein Auto zu bauen.

Dahinter stand welche Grundidee?

Wir wollten weg von einem hochkomplexen, hochintegrierten, über Jahre hinweg weiterentwickelten und ausgefeilten Gesamtprodukt, hin zu einem radikal einfachen Produkt. Ein Produkt, das sich für diesen Preis auch verkaufen lässt. Wir gingen von der Kernhypothese aus, dass die Elektromobilität für den Kunden ganz andere Aspekte relevant werden lässt.

Ein einfaches, simples Fahrzeug, das 5.000,– Euro kostet? Mussten Sie Ihr Ziel revidieren oder justieren?

Sicherlich was den konkreten Preis angeht. Der StreetScooter wird nicht für 5.000,– Euro

Where did the StreetScooter idea come from?

Christoph Deutskens: The idea arose at the institute. This was in 2008, 2009 when the issue of electromobility became more and more relevant. We discussed the impact these developments would have on the automotive industry. Even back then, it was clear that the entire automobile industry was changing very significantly and would therefore face massive upheavals. Initially, there was the topic of electromobility. At the same time, the question of how to bring research and practice closer together arose. This led us to the realization that we had to operate closer to the practice in order to advance an innovation ourselves. We needed a use case for this, a vehicle, to be able to demonstrate things. At this point, it was clear that it would be impossible for the electric vehicles to assert themselves as long as they were far more costly than conventional vehicles. This is how the idea came about: we wanted to develop an affordable electric vehicle.

What was the aim in the beginning?

The vision was to develop a car for Euro 5000.00.

Production cost?

The Euro 5000.00 had to be enough to build a car.

What was the basic idea?

We wanted to move away from a highly complex, highly integrated product, developed and refined over many years, to a radically simple product. A product that can also be sold for this price. We proceeded from the core hypothesis that the electromobility could result in quite different aspects relevant for the customer.

A basic, simple vehicle that costs Euro 5000.00? Did you need to revise or adjust your goal?

Certainly with regard to the specific price. The StreetScooter is not sold for Euro 5000.00.

verkauft. Es ist inzwischen auch ein etwas anderes Fahrzeug als ursprünglich entwickelt geworden, denn wir sind damals mit einem kleinen Zweisitzer gestartet. Aber grundsätzlich mussten wir an der Vision, ein bezahlbares, kostengünstiges Fahrzeug mit wenig Investition und geringer Komplexität zu entwickeln, nicht justieren.

Was wollten Sie sich und der Welt beweisen?

Wir wollten demonstrieren, dass es möglich ist, in kurzer Zeit, mit einem kleinen Team und wenig Investition ein Fahrzeug in Serie zu bringen. Etwas also das selbstverständlich niemandem in der Automobilindustrie selbstverständlich ist.

Warum nicht?

Wenn man betrachtet, welcher Aufwand heutzutage betrieben wird, um ein Fahrzeug in Serie zu bringen – da wurden wir anfangs belächelt. Niemand hat einen Erfolg für möglich gehalten. Es wurde zwar eingeräumt, dass wir den Forschungsanteil unserer Arbeit realisieren könnten, aber niemand ist davon ausgegangen, dass das Auto jemals in Serie gehen würde. Nicht mit einem solch kleinen Team. Man bräuchte Millionen, wenn nicht Milliarden an Invest – daher würde unser Vorhaben nicht gelingen.

Sie hatten also weniger mit Widerständen als mit Vorurteilen zu kämpfen?

Genau. In der Automobilindustrie sind alle Unternehmen groß, sehr etabliert und die meisten bereits jahrzehntelang aktiv. Darüber hinaus ist die Anzahl der Unternehmen begrenzt, sodass die Akteure sich untereinander kennen und eine Art Community bilden. In dieser Automobilindustrie wurden wir anfangs vielfach belächelt. Niemand dort hat daran geglaubt, dass unser Konzept aufgehen würde.

Das heißt, die Konzeptidee war immer öffentlich?

Ja, wir waren von Beginn an sehr offen, transparent, haben Leute eingeladen und

Since then, a slightly different vehicle than the original has been developed, since we started with a small two-seater. But basically, we did not have to adjust the vision of developing an affordable, low-cost vehicle with little investment and low complexity.

What did you want to prove to yourself and the world?

We wanted to demonstrate that it is possible to accomplish the serial production of a vehicle, within a short time, with a small team and little investment. Something that, of course, is not self-evident to anyone in the automotive industry.

Why not?

Considering the effort currently required to develop a vehicle – we were ridiculed at first. No one thought success was possible. It was recognized that we could realize the research share of our work, but no one assumed that the car would ever go into production. Not with such a small team. One would need millions, if not billions, of investment – so our project would not succeed.

So you were faced with less resistance than preconceptions?

Exactly. All companies in the automotive industry are large, very well established and most of them have been active for decades. In addition, the number of companies is limited so that the players know each other and form a kind of community. In this automobile industry, we were initially dismissed. No one there believed that our concept would bear fruit.

Does this mean that the concept idea was always public?

Yes, from the onset we have been very open and transparent. We invited and integrated people, and thus presented our concept publicly in many ways. For us as a university, this was always meant to be a demonstration case. Because among other things, we were dependent on partners. Due

integriert und so unser Konzept vielfach und intensiv nach außen dargestellt. Für uns als Hochschule sollte das immer ein Demonstration Case sein. Unter anderem aber auch, weil wir auf Partner angewiesen waren. Für die Kürze der Zeit, die wir uns vorgenommen hatten, war es wesentlich, mit vielen Partnern zusammenzuarbeiten. Und außerdem eine Frage der eigenen Philosophie: Wissen teilen führt nach unserer Ansicht zu einem Mehr an Wissen. Anstatt uns einzuschließen und der Welt nach drei Jahren zu zeigen, was wir gemacht haben, waren wir von vornherein an der Netzwerkidee orientiert. Das Konzept war in frühen Phasen sicherlich noch nicht serienreif, insofern hat unser Vorgehen an dieser Stelle sicherlich diese Vorurteile befördert. Derartiges abzubauen braucht bekanntermaßen seine Zeit. Im Endergebnis hat sich unsere Philosophie mit Sicherheit ausgezahlt.

Von wem haben Sie sich über die Schulter schauen lassen?

Anfangs haben wir mit Automobilzulieferern kooperiert, weniger mit den OEMs[44]. Mit der Zeit wurden aber einige Automobilhersteller auf uns aufmerksam und haben dann doch ein Interesse an einem Austausch und einer intensiveren Zusammenarbeit artikuliert.

Zu diesem Zeitpunkt hatten Sie aber schon einen Vorsprung?

Es existierten bereits erste vorzeigbare Ergebnisse.

Wie fühlten sich die Gespräche für die Zulieferer an?

Wir haben uns insbesondere zu Anfang auf Gespräche mit Automobilzulieferern und KMU[45] fokussiert. Das war für uns wesentlich, denn viele von diesen Ansprechpartnern waren motiviert, Leistung mit in das Projekt einzubringen. So konnten wir schneller Fahrt aufnehmen. Denen war es wichtig, nicht wie üblich jede Menge Restriktionen vorgesetzt zu bekommen. Im Gegenteil waren sie erfreut, zur Abwechslung mal durch höhere

to the time limitation involved, it was essential to work with many partners. Moreover, it was a question of our own philosophy: sharing knowledge leads, in our opinion, to a gain in knowledge. Instead of locking ourselves away and showing the world what we have achieved after three years, networking was a big focus for us from the get-go. The concept was certainly not yet ready for serial production at an early stage, so our approach at this point has certainly promoted these preconceptions. Dismantling this takes time, as we know. In the end, our philosophy has certainly paid off.

Who was able to come and look over your shoulders?

At the beginning, we cooperated with automotive suppliers and less with the OEMs[44]. Over time, however, some automotive manufacturers became aware of us and then expressed an interest in an exchange and an intensified cooperation.

At this point, you already have a head start, right?

There were already first presentable results.

How were the discussions with the suppliers?

At the beginning, we focused on discussions with automotive components suppliers and SMEs[45]. This was essential for us, because many of these respondents were motivated to contribute to the project. So we were able to move at a faster pace. It was important for them not to have a lot of restrictions as is customary. On the contrary, they were delighted to have had a great deal of freedom to develop their own solutions, for a change. We also consciously integrated many SMEs from the region. They often represent a somewhat casual approach. We had, however, SME or startup structures, in which not all processes are perfectly defined and where there are still no handbooks for everything. Under these circumstances, the cooperation with SMEs works better because the working methods are similar.

Freiheitsgrade eigene Lösungen entwickeln zu können. Wir haben zudem bewusst viele KMU aus der Region integriert. Die vertreten oft eine etwas hemdsärmelige Herangehensweise. Wir hatten allerdings selber KMU- oder Start-up-Strukturen, in denen nicht alle Prozesse perfekt definiert sind und es noch kein Handbook für alles gibt. Unter diesen Umständen gelingt die Zusammenarbeit mit KMU besser, weil die Arbeitsweisen sich ähneln.

Welche Probleme haben Sie zu Beginn einkalkuliert?

Es gab sicherlich ein paar Hürden, die von vornherein absehbar waren. Beispielsweise das Thema Homologation.

Was bedeutet Homologation?

Homologation bezeichnet den Prozess, durch den ein Produkt seine Straßenzulassung erhält. In den ersten Wochen unseres Projekts war uns nicht klar, wie wir das alles machen würden. Es gibt ein Standardprüfprogramm, das durchlaufen werden muss. Man muss beispielsweise nach Schweden, um das ABS[46] zu testen und zu programmieren. Man muss Crashtests absolvieren, mit der DEKRA[47] gewisse Prüfungen durchlaufen, bei denen die Sicherheitsfunktionen des Fahrzeuges untersucht werden. Gibt es ein Backup-System? Sind also beispielsweise gewisse Sicherheitsfunktionen doppelt ausgeführt? Erst dann erhält man die Straßenzulassung. Es gab also Herausforderungen, die uns von Anfang an bewusst waren. Während des Projekts gab es immer wieder Phasen, in denen man vor einem Berg stand und sich fragte, wie der zu meistern sei. Letztlich haben wir immer einen Weg gefunden. Tja, Probleme ansonsten ...?

Scheitern?

Mit dem Risiko des Scheiterns haben wir uns weniger befasst. Wir waren immer sehr selbstbewusst und motiviert und haben konstant an den Erfolg unseres Projekts geglaubt. Die Möglichkeit, das Projekt schon zu einem frühen Zeitpunkt zu beenden, gab es nicht. Das Ziel war klar: das Produkt in Serie zu bringen. Insofern gab es

46 ABS, Antiblockiersystem, technisches System für mehr Fahrsicherheit und weniger Verschleiß an den Laufflächen der Räder
ABS, anti-blocking system, technical system for more driving safety and less wear on the treads of the wheels

47 DEKRA, Deutscher Kraftfahrzeug-Überwachungsverein
DEKRA, German Motor Vehicle Supervisory Association

What problems did you make provisions for at the beginning?

There were certainly a few hurdles that were foreseeable from the outset. For example, the issue of homologation.

What does homologation mean?

Homologation means the process by which a product receives its approval to go on the road. In the first few weeks of our project, we did not understand how we would accomplish everything. There is a standard test program that has to be performed. For example, you are required to go to Sweden to test and program the ABS[46]. Crash tests must be carried out, certain tests must be performed with DEKRA[47], in which the safety functions of the vehicle are examined. Is there a backup system? Are, for example, certain safety functions duplicated? Only then can you get the road approval. So there were challenges that we were aware of from the beginning. During the project, there were always times when you were standing in front of a mountain and wondering how to master it. Ultimately, we have always found a way. Well, problems, apart from that ...?

Failure?

We were less concerned about the risk of failure. We were always very self-confident and motivated and have consistently believed in the success of our project. There was no option to terminate the project at an early stage. The goal was clear: the serial production of the product. In this respect, there was also no plan B to which we would have limited ourselves, for example, to the development of a research vehicle. It was more a matter of technical risks, such as homologation. Because it is like this: at the moment that we offer the vehicle, we bear a responsibility and a risk. That means that it is relevant to adequately cover all necessary safety functions. These are the risks on which we concentrated. Moreover, we had to pay attention to the budget that was always very limited during the entire process.

auch keinen Plan B, in dem wir uns beispielsweise auf die Entwicklung eines Forschungsfahrzeugs beschränkt hätten. Es handelte sich eher um technische Risiken, also Themen wie die Homologation. Denn es ist ja so: In dem Moment, in dem wir das Fahrzeug anbieten, tragen wir eine Verantwortung und ein Risiko. Das heißt, es ist relevant, alle nötigen Sicherheitsfunktionen vernünftig abzudecken. Das sind die Risiken, auf die wir uns konzentriert haben. Außerdem mussten wir permanent auf das Budget achten, das während des gesamten Prozesses immer sehr limitiert war.

War das Projekt „StreetScooter" von vornherein eine Ausgründung oder wie lief das formal ab?

Der Prozess lief auf formaler Ebene folgendermaßen: Wir sind 2009 als Hochschulprojekt gestartet. Damals hatten sich mehrere Hochschulinstitute mit dem Ziel zusammengeschlossen, gemeinsam ein Fahrzeug zu entwickeln. Dann wurden Aufgaben verteilt. Das ging bis 2010. Im Kern haben wir dann in drei, vier Monaten, also relativ schnell, ein Fahrzeugkonzept entwickelt und ausgearbeitet, mit dem wir dann an die Unternehmen herangetreten sind, um sie für unser Projekt zu begeistern. Daraus sind verschiedene Kooperationsformen entstanden, einige haben entweder in die geplante StreetScooter GmbH investiert, andere haben eine gewisse Eigenleistung im Rahmen der Entwicklung übernommen. Offiziell haben wir die StreetScooter GmbH als Firma Mitte 2010 gegründet.

Wie wurde das Team rekrutiert?

Das Kernteam bestand zunächst aus wissenschaftlichen Mitarbeitern verschiedener Institute. Als die StreetScooter GmbH dann gegründet war, hat man auch Festangestellte übernommen. Das waren zum einen Uni-Absolventen, von denen man wusste, dass sie im Rahmen ihrer Master- oder Diplomarbeiten schon am Thema StreetScooter mitgearbeitet hatten. Andererseits wurden Mitarbeiter extern aus Firmen der Region rekrutiert, die uns mit ihren Erfahrungen unterstützen konnten. Das

Was the "StreetScooter" project a spin-off from the beginning or what were the formalities?

The formal process was as follows: we started in 2009 as a university project. At that time, several university institutes joined to develop a vehicle together. Then tasks were distributed. This went on until 2010. In essence, we then developed and worked out a vehicle concept in three or four months, that is relatively quick, with which we then approached the companies to get companies enthused about our project. Various forms of cooperation emerged from this, some of which were invested in the planned StreetScooter GmbH, others have made a certain amount of their own contribution to the development. Officially, the StreetScooter GmbH as a company was established in mid 2010.

How was the team recruited?

The core team initially consisted of scientific staff from various institutes. When the StreetScooter GmbH was founded, we recruited permanent employees. These included, on the one hand, university graduates, who were known to have worked on the subject of StreetScooter as part of their Masters or Diploma thesis. On the other hand, employees were recruited externally from companies in the region who were able to support us with their experience. This was done by word of mouth, that is, someone knew someone whom he was able to motivate to make the move. Many employees switched from their positions to StreetScooter GmbH because they were convinced of the idea and the company.

What did the particular moment when this idea emerged feel like? Was it an idea conceived by two people over a beer?

From a sober perspective – it was actually unspectacular. As far as I know, this was in Professor Schuh's office, along with Professor Kampker and Tobias Reil, the current Head of Production & IT at the company StreetScooter.

lief meistens auf einer persönlichen Schiene, das heißt, jemand kannte jemanden, den er zu einem Wechsel motivieren konnte. Viele Mitarbeiter sind also aus ihren Positionen heraus zu StreetScooter GmbH gewechselt, weil sie von der Idee und dem Unternehmen überzeugt waren.

Wie kann man sich den Entstehungsmoment dieser Idee vorstellen? Haben da zwei Köpfe bei einem Bier gesessen?

Nüchtern betrachtet – unspektakulär eigentlich. Meines Wissens war das damals im Büro von Professor Schuh, gemeinsam mit Professor Kampker und Tobias Reil, dem heutigen Head of Production & IT der Firma StreetScooter.

Die Männer der ersten Stunde? Aber ohne Bier.

Ich war persönlich ja leider nicht dabei. Ich weiß aber, dass Professor Kampker gemeinsam mit Tobias Reil ein erstes Team definiert hat. Ich kam kurze Zeit später dazu. Als Mann der zweiten Stunde. Wir hatten damals jeden Freitag einen fixen Termin. In einem kleinen Besprechungsraum kam das Team zusammen.

Mit Blick auf den Industrialisierungsprozess – sehen Sie sich als Pionier oder als Evoluzzer, der etwas weiterentwickelt?

Wenn man den gesamten Prozess bis zur Serienproduktion betrachtet, ist das revolutionär. In der Radikalität, mit tatsächlich so wenig Invest, in der Kürze der Zeit und dem, was wir da alles geschaffen haben, ist das schon revolutionär. Zumindest in der Branche, in dem Umfeld, wüsste ich kein vergleichbares Beispiel. Wenn man auf die Methodik schaut und die Details betrachtet, ist sicherlich nicht jeder Aspekt revolutionär. Wir haben uns selbstverständlich von manchen Dingen inspirieren lassen. Es gibt Andere, die zum Teil zumindest theoretisch ein derartiges Vorgehen vorschlagen. Aber das Ergebnis und die Idee von Return on Engineering würde ich als revolutionär bezeichnen.

The men of the first hour? But without beer.

I was unfortunately not there. I know, however, that Professor Kampker, along with Tobias Reil set up a first team. I came on board a short time later as a man of the second hour. We had a fixed meeting every Friday. The team came together in a small meeting room.

Overlooking the industrialization process – do you see yourself as a pioneer or an evolutionary who is developing something further?

Looking at the entire process, right up to serial production, it is revolutionary. In the radicalism, with so little investment, within the short time, and everything we have created, it is really revolutionary. I would not have a comparable example at least in this industry or in this environment. Looking at the methodology and at the details, certainly not every aspect is revolutionary. We were of course inspired by some things. There are others which, in part, suggest such an approach, at least theoretically. But I would describe the result and the idea of the Return on Engineering as revolutionary.

Eine Idee in die Tat umsetzen:
Prof. Achim Kampker, CEO
StreetScooter GmbH und
Bereichsleiter Elektromobilität
bei Deutsche Post
Turning an idea into reality:
Prof. Achim Kampker, CEO
StreetScooter GmbH and Head of
Electromobility at Deutsche Post

QUALITÄT UND INNOVATION KANN AUCH BEDEUTEN,

„"

Prof. Achim Kampker
Geschäftsführer der Streetscooter GmbH
Executive Vice President E-Mobilität Deutsche Post DHL
CEO of Streetscooter GmbH
Executive Vice President E-Mobility Deutsche Post DHL

EINFACHES MIT EINFACHEN MITTELN

ZU LÖSEN.

8

DIE STREETSCOOTER-PRODUKTIONS-STORY

THE STREETSCOOTER PRODUCTION STORY

Die Mobilitätswende ist ein Megathema unserer Zeit. Der Wandel vom konventionellen zum elektrischen Antriebsstrang betrifft die gesamte Wertschöpfungskette von Automobilherstellern über Komponentenlieferanten bis hin zum Maschinen- und Anlagenbauer. Innovative Produktionsprozesse und Fahrzeugkonzepte spielen dabei eine zentrale Rolle.

The mobility revolution is a big issue of our time. The change from conventional to electric drivetrains affects the entire value chain, from automobile manufacturers to component suppliers back to the machine and plant manufacturers. And here innovative production processes and vehicle concepts play a central role.

INNOVATION, PRODUKTION, ÖKOVISION

INNOVATION, PRODUCTION, ECO VISION

Innovative Fahrzeugentwicklung

Aha, ein Elektroautomobil! Überraschung? Nein, das allein ist nicht das Einzigartige. Das Besondere an StreetScooter ist eben nicht nur die Neukonzeption eines serienreifen und günstigen Elektromobils und spezieller Komponenten, sondern der völlig neue Blick auf das gesamte Thema der Fahrzeugentwicklung. Es geht dabei um die ganzheitliche Betrachtung von Produktentwicklung und Produktionsprozessen. Diese umfassende Sichtweise stellt einen entscheidenden Faktor für die angestrebte wirtschaftliche Lösung dar. Die Entscheidungen, die bereits während der Produktentwicklung getroffen werden, machen später rund 80 % der Gesamtkosten aus. Das Ergebnis der ganzheitlichen Betrachtung des Produktionsprozesses ist eine modulare Fahrzeugarchitektur, mit der unterschiedliche Varianten elektrisch betriebener Fahrzeugtypen einfach und wirtschaftlich realisiert werden können. Die Prototypen-Präsentation bei potenziellen Kunden ist innerhalb eines Jahres möglich – ab Festlegung der gewünschten Anforderungen und inklusive eines kundenspezifischen Testings. StreetScooter stellt mit seinen Entwicklungs- und Produktionsprozessen einen neuartigen Ansatz der Automobilentwicklung und -produktion dar, insbesondere im Hinblick auf die Elektromobilität. Losgelöst von bisherigen Denkstrukturen und -verfahren zeigt sich, dass nicht nur ein Elektrofahrzeug schnell, wirtschaftlich und nutzenspezifisch entwickelt werden kann, sondern auch Kleinserien von Elektrofahrzeugen

Innovative vehicle development

Wow, an electric car! Surprised? No, that's not what makes it unique. What makes StreetScooter special is not a new conception of a series production-ready and cheap electric vehicle and special components, but the completely new way of looking at the whole issue of vehicle development. In other words, a holistic approach to product development and production processes. This comprehensive approach represents a decisive factor for achieving the desired economic solution. The decisions that are already made during product development will later translate to nearly 80 % of the total cost. The result of the holistic view of the production process is a modular vehicle architecture, with which different versions of electrically powered vehicle types can be realized easily and economically. The prototype presentation to potential customers can happen within a year – after defining the desired requirements and including a customized testing. StreetScooter's development and production processes represent a novel approach to automobile development and production, especially with regard to electromobility. Detachment from previous ways of thinking and methods shows that not only can an electric vehicle be quickly, economically, and use-specifically designed, but also small series of electric vehicles can be produced economically. In this respect, StreetScooter represents a quasi-disruptive element in the automotive industry and enables hitherto non-viable electric vehicle solutions, especially for fleet operators.

StreetScooter-
Innenansichten:
Der „Work" entsteht
in Progress
*StreetScooter interior
views: The "Work" is
created "as you go"*

Mathias Kammüller
Geschäftsführer TRUMPF GmbH & Co. KG
Managing Director TRUMPF GmbH & Co. KG

„Der StreetScooter ist mit seiner neuartigen Karosseriefertigung auch für TRUMPF ein sehr spannendes Projekt. Wir haben dafür gemeinsam mit der RWTH Aachen ein neues Konstruktionsprinzip entworfen, mit dem der komplette Unterboden des Fahrzeugs Laser-gerecht konstruiert wird. Die flachen Blechteile werden auf unseren Laser-Maschinen in Form geschnitten, einfach zusammengesteckt und mit dem Laser punktuell verschweißt. Das spart Material – und somit auch Gewicht und Energie. Durch die neue Konstruktion des Unterbodens fallen pro Schweißkante mehrere Millimeter Blech weniger an. In der Serienproduktion bedeutet das unter dem Strich aber eine Einsparung von Hunderten Tonnen an Stahl."

"The StreetScooter with its new car body manufacturing is also a very exciting project for TRUMPF. Together with the RWTH Aachen, we designed a new construction principle where the entire underbody of the car is built using lasers. The flat sheet metal parts are cut into shape on our laser machines, then simply put together and welded at selected points. This saves material – and thus weight and energy, too. With the new design of the underbody, there are several millimeters less plate metal per welding edge. But the bottom line is that in series production this translates into savings of hundreds of tons of steel."

wirtschaftlich produziert werden können. Insofern stellt StreetScooter ein quasi-disruptives Element in der Automobilwirtschaft dar und ermöglicht bislang nicht realisierbare Elektrofahrzeuglösungen, speziell für Flottenbetreiber.

Produktive Effizienz

Die Ingenieure des StreetScooters hatten anfangs mit drei großen Herausforderungen kreativ lösungsorientiert umzugehen: 1. Die Herstellung der ersten Vorserie des Fahrzeugs musste mit einem Team von weniger als dreißig Personen gestemmt werden. 2. Die Montagelinie für 2.000 Fahrzeuge pro Jahr war mit weniger als 1 Mio. Euro zu finanzieren. 3. Mit dem Entwicklungsauftrag der Post musste die Produktion von 0 auf 200 Fahrzeuge innerhalb von zweieinhalb Jahren hochgefahren werden, inklusive aller technischen Zulassungen. Die StreetScooter GmbH hat diese Phasen zwischen 2010 und 2012 erfolgreich durchlaufen. Inzwischen haben die Aachener

Productive efficiency

The engineers of the StreetScooter had to first deal with three major challenges creatively and pragmatically: 1. Menufacturing of the first pilot production of the vehicle had to be accomplished with a team of fewer than thirty people. 2. The assembly line for 2,000 vehicles per year was to be financed with less than 1 million euros. 3. With the development contract from Deutsche Post, production had to be boosted from zero to 200 vehicles within two and a half years, including all technical approvals. StreetScooter GmbH successfully went through these stages between 2010 and 2012. Meanwhile, the Aacheners developed several models and for the most part brought them to the production stage: from the electric bike to the compact city cars to light commercial vehicles – such as the delivery vehicles for the Deutsche Post DHL Group. From the initial idea of making electromobility economically attractive in small volumes and of combining economy and ecology, a successful example of productive efficiency and affordable

mehrere Modelle entwickelt und zum größten Teil zur Serienreife gebracht: vom Elektrofahrrad über den kompakten Stadtflitzer bis zu leichten Nutzfahrzeugen – wie zum Beispiel den Zustellfahrzeugen für die Deutsche Post DHL Group. Aus der anfänglichen Idee, Elektromobilität bereits ab kleinen Stückzahlen wirtschaftlich attraktiv zu gestalten und Ökonomie und Ökologie miteinander zu verbinden, wurde ein erfolgreiches Beispiel für produktive Effizienz und bezahlbare Funktionalität. Im Mittelpunkt des Ansatzes stehen die Halbierung der Entwicklungszeit, die Reduzierung der Investitionskosten um bis zu 90 % sowie eine modulare Fahrzeugarchitektur. Darauf basierend wurde ein auf die individuellen Bedürfnisse der Deutsche Post DHL Group ausgelegtes Elektrofahrzeug namens „Work" entwickelt. Dieses in enger Kooperation mit Zustellern der Post entwickelte Nutzfahrzeug und spezieller Komponenten wurde 2012 präsentiert und ist seit 2014 im bundesweiten Flottentest bei der Deutschen Post DHL Group. Die Serienproduktion des „Work" und der Komponenten ist in 2015 angelaufen. Das Konzept und der Erfolg führten dazu, dass im Dezember 2014 die StreetScooter GmbH zu 100 % durch die Deutsche Post DHL Group übernommen wurde. Am Standort Aachen werden heute Elektrofahrzeuge für den Kurzstreckeneinsatz entwickelt und produziert. Im Fokus stehen dabei Fahrzeuglösungen für die Zustellung auf der sogenannten „letzten Meile". Kommunale Einrichtungen, Logistikdienstleister sowie andere Unternehmen setzen auf StreetScooter im Rahmen ihrer Flottenlösungen.

functionality came into being. At the center of the approach is cutting the development time in half, reducing investment costs by up to 90 %, and a modular vehicle architecture. It was on the basis of this approach that the electric vehicle called "Work" was developed, geared to the individual needs of the Deutsche Post DHL Group. This commercial vehicle and special components, which was developed in close cooperation with Deutsche Post carriers, was presented in 2012 and since 2014 has been in a nationwide fleet test with the Deutsche Post DHL Group. Series production of the "Work" and its components began in 2015. The concept and the success led Deutsche Post DHL Group's 100 % acquisition of StreetScooter GmbH in 2014. Today, at the Aachen site, electric vehicles for short-range use are developed and produced. The focus here is on vehicle solutions for delivery along the so-called "last mile". Communal facilities, logistics service providers, and other companies rely on StreetScooter as part of their fleet solutions.

Environmentally friendly cost efficiency

Electromobility makes a significant contribution to the reduction of emissions such as CO_2, fine dust and noise. However, even the conservation of resources is only a positive feature provided that the necessary driving energy is produced in a carbon neutral way. The reconciliation of economy and ecology, today the guiding principle

Umweltfreundliche Wirtschaftlichkeit

Elektromobilität leistet einen wesentlichen Beitrag zur Reduzierung von Emissionen wie CO_2, Feinstaub und Lärm. Aber auch die Ressourcenschonung fällt positiv ins Gewicht, vorausgesetzt, die notwendige Antriebsenergie wird klimaneutral hergestellt. Die Versöhnung von Ökonomie und Ökologie, heute der Leitgedanke der StreetScooter GmbH, war zu Beginn der Entwicklung des StreetScooters eine große Herausforderung. Ökonomie bei kleinen Stückzahlen und klassischerweise hohen Industrialisierungsaufwänden passte zunächst nicht zusammen.

Mit dem StreetScooter wurden neue Wege beschritten, um einen höheren Return on Engineering zu erzielen. Auch die Produktion des StreetScooters sieht heute anders aus als eine klassische Automobilproduktion. StreetScooter produziert auf einem alten Industriegelände in Aachen. Auf der aktuellen Montagelinie können bis zu 10.000 Fahrzeuge pro Jahr produziert werden. Rund 100 Mitarbeiter sind in der Montage beschäftigt und produzieren aktuell im Ein-Schicht-Betrieb. Einfache Technologien mit niedrigen Investitionen sowie clevere Hilfsmittel und Vorrichtungen prägen das Bild. Neben der Montagelinie befinden sich verschiedene Vormontagebereiche und Logistikflächen auf dem Gelände. Und vor allem gibt es genügend Platz zum Wachsen. Schon jetzt werden weitere Hallenflächen erschlossen und an der Erweiterung der Produktion gearbeitet.

of StreetScooter GmbH, was a big challenge at the beginning of the development of StreetScooter. Initially, economy with small production quantities and traditionally high industrialization costs did not go together.

StreetScooter explored new avenues for achieving a higher Return on Engineering. Even the production of StreetScooter looks different than a classic automobile production. StreetScooter produces at an old industrial site in Aachen. Up to 10,000 vehicles can be produced per year on the current assembly line. Around 100 people work in assembly and are currently producing in a one-shift operation. Simple technologies with low investments and clever tools and devices feature prominently here. In addition to the assembly line, there are several pre-assembly areas and logistics facilities on the premises. And above all, there is plenty of room to grow. Already, more warehouse space has been opened up and work is underway on expanding the production area.

Alles fließt: von der
Montagestation bis
zur Ladestation
*Everything flows
seamlessly: From the
assembly station to
the loading station*

Tobias Reil
Head of Production & IT Aachen, StreetScooter GmbH, Aachen
Head of Production & IT Aachen, StreetScooter GmbH, Aachen

Wie muss man sich Ihr Tätigkeitsfeld vorstellen?

Tobias Reil: Mein Aufgabenbereich bei der StreetScooter GmbH erstreckt sich über das gesamte Thema der Produktion, über das komplette Produkt-Portfolio hinweg, vom kleinen Nutzfahrzeug über das große Nutzfahrzeug bis zum Zwei- und Dreirad – damit verbunden sind auch Ausbau und Erweiterung der gesamten Wertschöpfungstiefe zu sehen.

Sie sind ein Spross der Hochschule?

Ja, ich bin Maschinenbauer und Wirtschaftsingenieur.

Sie laufen also ständig mit dem Rotstift durch die Produktion?

Na ja, das bringen Studium und meine Aufgabe hier schon mit. Wobei es nicht darum geht, Dinge wegzusparen, sondern immer zu sehen, was kann man noch verbessern und effizienter gestalten. Wirkung plus Wirtschaftlichkeit.

Und wohin geht da die Reise?

Ich denke, wir haben noch viele Themen und viele Baustellen, die wir in den nächsten Jahren unternehmerisch meistern müssen. Aber das kennen wir nicht anders. Hatten wir bisher auch immer. Und genauso haben wir bisher immer alles bravourös geschafft. Ich denke, das ist erst einmal unsere Aufgabe. Wir fahren jetzt die Stückzahlen hoch und das Hochfahren der Stückzahl bringt entsprechend neue Aufgaben und Herausforderungen mit sich, die wir kontinuierlich weiter fixen müssen. Wir werden in den nächsten Jahren vielleicht auch das Produktprogramm in Maßen weiter ausrollen.

Sie werden aber nicht auf den privaten Sektor gehen, oder?

Wir werden zunächst weiter auf den Flottenbetreiber zielen. Aber wer weiß, wo wir in fünf Jahren stehen.

Flotte heißt heute Deutsche Post?

Heißt heute Post. Das können aber auch städtische Fuhrparkbetreiber sein, die eine Stückzahl

What is your responsibility?

Tobias Reil: My area of responsibility at StreetScooter GmbH extends to the entire production line, across the entire product portfolio, from the small commercial vehicle to the large commercial vehicle to the two- and three-wheeler, which also includes the expansion and extension of the entire depth of the added value.

Are you an alumnus of the university?

Yes, I am a mechanical and industrial engineer.

So you are always running through the production, red pencil in hand?

Well, that is in fact a part of my studies and my responsibility here. It is not a matter of saving things, but always seeing what can be improved on and made more efficient. Impact plus economic efficiency.

And where is the journey leading?

I think we still have many issues and a lot of roadblocks that we have to overcome in the coming years. But we do not know it any other way. We have always had those. And so far we have always accomplished everything brilliantly. I think that is our initial task. We are now increasing the number of units and an increase in the number of units comes with new tasks and challenges, which we must continually fix. In the coming years, we are also likely to continue moderately rolling out the product range.

But you will not target the private sector, right?

We will continue to target the fleet operators. But who knows where we will be in five years.

Fleet today means Deutsche Post?

Currently it is the Post. However, it can also be urban car fleet operators which use at least 50 vehicles in the local metropolitan areas.

von mindestens 50 Fahrzeugen in den lokalen Ballungsräumen in den Einsatz bringen.

Ab wann waren Sie beim Thema StreetScooter eigentlich dabei?

Stunde null.

Initiator?

Nein, der Anstoß kam von Professor Schuh.

Einer guten Idee ist es ja egal, wer sie hatte.

Sehe ich auch so. Hauptsache, jemand stößt eine Entwicklung an und hat eine Idee oder stellt sich eine Frage, die nach einer Beantwortung verlangt.

Welche Frage stand denn in den Kindertagen des StreetScooters im Raum?

Zuerst war da noch kein StreetScooter, aber natürlich gab es eine Fragestellung, die letztendlich zum StreetScooter führte. Die Frage war, ob es mit den heutigen Technologien möglich sein würde, Elektromobilität in die breite Öffentlichkeit zu bringen. Warum sollte das nicht gehen? Zu der Zeit, also etwa 2009, schwebte immer dieses Tesla-Beispiel über allem, wenn man von E-Mobilität sprach. Da waren in Kalifornien die ersten Elektrofahrzeuge für einen Preis von um die 100.000 bis 120.000 Euro auf der Straße aufgetaucht, auch teils erfolgreich angenommen worden.

Aber eher von wenigen mit dem nötigen Kleingeld.

Ja, eben nichts für die Masse. Aber das war ja genau unser Punkt. Wir wollten zeigen, dass es auch günstiger geht, deutlich günstiger. Auch mit kleineren Stückzahlen.

In der Herstellung und im Verkauf günstiger?

Ja, genau. Daraufhin sind dann an der Hochschule etwa 18 oder 20 Institute, die alle irgendwas mit dem Thema Mobilität zu tun haben, zusammengetrommelt worden; also Kunststofftechnik, Fahrzeugtechnik, Stahlbau, Produktionstechnologie und so weiter und so fort.

Since when have you actually been with StreetScooter?

Zero hour.

Initiator?

No, the impulse came from Professor Schuh.

It is a good idea regardless of who came up with it.

I agree with you. The main idea is that someone initiates a development, and has an idea, or that a question arises requiring an answer.

What was the question in the infancy stages of the StreetScooter?

At first, there was no StreetScooter, but of course there was a question which ultimately led to the StreetScooter. The question was whether or not, with today's technologies, it would be possible to offer electromobility to the general public. Why shouldn't that work? At the time, in about 2009, this Tesla example hovered over everything when the topic of e-mobility came up. In California, the first electric vehicles were already on the road for a price of around Euro 100,000 to Euro 120,000 and were also partially accepted.

But only by a few with the necessary small change.

Yes, nothing for the general public. But that was precisely our point. We wanted to show that it could also be more affordable, significantly more affordable. Also with smaller numbers.

More affordable in production and sale?

Yes exactly. As a result about 18 or 20 institutes, all of which deal with the topic of mobility, were rounded up at the university; including plastics technology, vehicle technology, steel construction, production technology and so on and so forth.

Wer hat getrommelt?

Professor Schuh. Auf dieser Basis haben eben auch diese 18 oder 20 Professoren, die grundsätzlich die Idee alle gut fanden, ihr Commitment gegeben. Und dann ist daraus erst mal eine professorale Projektgruppe entstanden und die Geschäftsstelle Elektromobilität ins Leben gerufen worden. Mit die erste Erkenntnis war dann: Wir können vieles, aber eben nicht alles. Wir brauchen auch ein starkes Industrie-Konsortium an unserer Seite.

Das dann erfolgreich akquiriert wurde?

Es ist ein Industrie-Konsortium aufgebaut worden. Letzten Endes, so nach einem Jahr, hatten sich zwischen 50 und 75 Industriepartner gefunden, die sich bereit erklärt hatten, das Projekt mit zu unterstützen. 2010 wurde dann die StreetScooter GmbH gegründet.

Weg vom Unternehmen, hin zum Produkt: War eigentlich das Design beim StreetScooter nie ein Thema?

Es gab einen externen Designer ...

... der hat sich hingesetzt und einen Stift genommen und gesagt: So muss das Produkt aussehen?

Design stand nicht im Vordergrund, sondern Design to Cost. Was der Designer an Gestaltung vorschlug, musste immer auch unter der Low-Cost-Prämisse realisiert werden können. Bei unserer Art der Produktentwicklung haben wir konsequent für einzelne Komponenten immer die kostengünstigste Lösung in der Entwicklung gesucht. Wir haben wie bei allem so auch innerhalb der Produktgestaltung stets Alternativen in Betracht gezogen und so immer kostengünstigere, aber nach unserer Einschätzung auch gleichwertige Lösungen ermittelt.

Das war die grundsätzliche Herangehensweise? Design to Cost?

Ja, sicher. Wir hatten im Konsortium zum Beispiel Kunststoffspezialisten und Werkzeugspezialisten, die immer rechtzeitig sagen konnten: Okay, so etwas kann man zwar produzieren, wird aber sehr teuer.

Who rounded them up?

Professor Schuh. On this basis, these 18 or 20 professors, who basically also found the idea to be a good one, gave their commitment. And then, a professors' project group was created and the electromobility business unit was created. The first realization was then: we can do a lot, but not everything. We also needed a strong industry consortium by our side.

Which was then successfully acquired?

An industry consortium was established. In the end, after a year, between 50 and 75 industrial partners, who had agreed to support the project, were found. In 2010, the StreetScooter GmbH was founded.

Away from the company, towards the product: Was the design of the StreetScooter never a topic?

There was an external designer ...

... who took pen to paper and said: This is how the product should look?

Design was not at the forefront, but design to cost. Whatever the designer proposed to design, had to be realized under the low-cost premise. In our way of product development, we have consistently looked for the cost-effective but equivalent quality solution for individual components. As always, we have always considered alternatives, including in the product design, and we have always found more cost-effective and, in our opinion, equivalent solutions.

That was the basic approach? Design to cost?

Yes, certainly. The consortium consisted, for example, of plastic specialists and tool specialists, who could always say in time: okay, something like this can be produced, but it would be very expensive. If you do it this way or that, it may not look so chic, but it would be produced more cost-effectively. Everything was questioned, why, for what reason does it have to be this way? Is this really necessary? Will this really bring the added

Wenn man das so und so macht, dann sieht es vielleicht nicht ganz so chic aus, aber dafür kann man es billiger herstellen. Es ist immer alles hinterfragt worden, wieso, weshalb, warum muss das so sein? Ist das wirklich notwendig? Bringt das wirklich nachher den Mehrwert oder eben nicht? Wenn die Kosten der Funktionalität nicht durch den Nutzen übertroffen wurden, wurden ganz klare Entscheidungen gegen die teurere Variante getroffen.

Können Sie ein Beispiel geben?

Nehmen wir einen Weltmarktführer für Schließsysteme, der fast alle Automobilzulieferer mit Funkschließsystemen ausrüstet. Die Anforderung der Automobilindustrie hinsichtlich der Reaktionsschnelligkeit des Schlosses ist die: Wenn ich auf den Knopf drücke, muss das Schloss am Auto innerhalb von 0,03 Sekunden öffnen. Ein solches Schlosssystem kostet dann entsprechend einen höheren Betrag X. Diese führende Firma für Funkschließsysteme hat aber auch ein eigenes System entwickelt, das ungefähr 40 bis 50 % billiger ist. Und da liegt die Reaktionszeit bei 0,06 Sekunden. Das sind dann die Beispiele, die wir gerne aufgegriffen haben, weil uns klar war: Das macht keinen Unterschied, das merkt keiner. Das heißt, die Funktionalität, dass das Öffnen der Türe 100 % länger dauert, wird der Kunde nicht feststellen. Für uns keine Frage, dass wir diesen Kompromiss eingingen. Dafür haben wir ein Schließsystem bekommen, das nur 50 % dessen kostet, was es sonst kostet.

Da gibt es ja sicherlich noch mehr Beispiele bezüglich der Komponenten, die minimal weniger Qualität liefern, aber deutlich günstiger sind.

Ja, sicher. Bei den Automobilzulieferern liegt das Potenzial zum Teil im Regal. Man muss die Komponenten nur ernsthaft in Betracht ziehen.

Lassen wir das mal so im Raum stehen – greifen wir noch einmal den StreetScooter auf und schauen auf die Produktion. Wo sind denn aktuell die Herausforderungen zu sehen?

value in the end or not? If the cost of the functionality was not exceeded by the benefits, clear decisions were made against the expensive variant.

Can you give an example?

Let us take a global market leader for locking systems, who supplies almost all automotive component supplies with remote keyless entry (RKE) systems. The requirement of the automotive industry with regard to the response speed of the lock is: when I press the button, the lock on the car must open within 0.03 seconds. Such a locking system costs a corresponding amount of X. This leading company for RKE systems has also developed its own system, which is about 40 to 50 % cheaper. And the reaction time for this is 0.06 seconds. These are the examples that we liked to take advantage of, because we realized that this does not make a difference, no one notices this. That is, the functionality that the opening of the door takes 100 % longer, will not be noticed by the customer. For us, there was no question: we accepted this compromise. And now we have a locking system that costs only 50 % of what it would cost otherwise.

There are certainly still more examples concerning components, which deliver minimally less quality but are significantly cheaper.

Yes, certainly. For the automotive components suppliers, the potential is partially on the shelf. You only have to seriously consider the components.

Let's leave that as it is – let's again look at the StreetScooter production. What are the challenges currently?

Technically, there are always issues that are not serious and can always be solved relatively quickly.

Where do the parts come from? Germany? Europe?

Technisch gibt es immer wieder Themen, die aber nicht gravierend sind und immer relativ schnell gelöst werden können.

Woher kommen die Teile? Deutschland? Europa?

90 % der Lieferanten sind in Deutschland, ein paar Prozent kommen aus dem näheren europäischen Umfeld. Wenige Komponenten aus den USA.

Bekommen Sie Feedback oder Kritik aus dem Anwenderbereich und werden die Hinweise verarbeitet?

Wir sind sehr nah an der Praxis. Wir haben unsere Mitarbeiter im Feld. Wir bekommen automatisch Rückmeldungen, tägliche Reports oder auch direkte Mitteilungen von den Zustellern der Post, wenn irgendwas nicht funktioniert, darunter sind auch Verbesserungsvorschläge, die aus dem Anwendungsfeld kommen, die dann entsprechend bewertet werden und ggfs. in die Produktion oder auch in die Weiterentwicklung eingesteuert werden.

Abschließend eine Frage an Sie persönlich: Was hat Sie motiviert, an der StreetScooter-Story mitzuschreiben?

Ich wollte ganz einfach etwas bewegen. Das Thema Elektromobilität an sich ist schon spannend. Aber es ging mir auch darum, zu zeigen, dass wir mit einem radikal anderen Denkansatz hinsichtlich Entwicklung und Realisation auch andere Wege beschreiten können. Das konnte ich nur in einer solchen Konstellation aus exzellenter Forschung und engagiertem, mittelständischem Unternehmertum, also diesem flexiblen Industrie-Konsortium, das hier mitgezogen hat. Meine Entscheidung war richtig. Und man sieht es ja auch: Die Firma wächst weiter. Die nächste Erfolgsgeschichte wird darin bestehen, zu zeigen, wie wir die Herausforderungen des Wachstums meistern werden.

Vom Start-up zum Grow-up. Aber Sie bleiben am Standort Aachen?

Wir haben hier vor ein paar Wochen gerade in eine neue Produktionshalle investiert.

90 % of the suppliers are in Germany and a few percent come from the immediate European regions. Few components are from the USA.

Do you receive feedback or criticism from the user segment and are these worked on?

We are very close to the market. We have our employees in the field. We automatically receive feedback, daily reports or direct messages from the postal service providers, if something does not work. This also includes suggestions for improvement that come from the field of application, which are then evaluated accordingly and, if necessary, are put into production or even in enhancement.

Finally, a question to you personally: What motivated you to become involved in writing the StreetScooter story?

I just wanted to make a difference. The topic of electric mobility itself is already exciting. But I also wanted to show that, with a radically different approach to thinking in terms of development and realization, we could also follow other paths. I would have only been able to do this as a part of such a constellation of excellent research and committed, medium-sized enterprises, meaning this flexible industry consortium that has been involved in this. My decision was the right one. And you also see it: the company continues to grow. The next success story will be to show how we will master the challenges of growth.

From start-up to grow-up. But you will remain at the Aachen location?

We have just invested in a new production hall here a few weeks ago.

WIR SEHEN UNS ALS VORREITER UND INNOVATIONSFÜHRER

UNTER DEN
BETREIBERN
KOMMERZIELLER
FAHRZEUGFLOTTEN.

Jürgen Gerdes
Mitglied des Vostandes Deutsche Post AG
Member of the board Deutsche Post AG

9

DIE STREETSCOOTER-POST-STORY

THE STREETSCOOTER POST STORY

Gelb fährt vor! Wie man bei einer Entwicklung ganz neue Wege beschreitet und wie dabei die (Deutsche) Post abgeht: Vom Feldversuch-Partner bis zur Übernahme der StreetScooter GmbH.

Yellow takes the lead! How we broke new ground with a development and how things took off like a (Deutsche Post) train: From field test partner to acquisition of StreetScooter GmbH.

NICHT NUR BRIEFE, SONDERN

IDEEN TRANSPORTIEREN.

Umwelt- und zustellerfreundlich:
Die Post will in den kommenden
Jahren rund 30.000 Fahrzeuge
in ihrem Fuhrpark ersetzen.
Environmental and deliverer friendly:
Deutsche Post looking to replace
approximately 30,000 vehicles in
its fleet in the coming years.

Bei der Deutschen Post in Bonn hatte von einer Firma namens StreetScooter bis zu jenen Tagen im September 2011 noch niemand gehört. Das mag zum einen daran gelegen haben, dass der Street-Scooter elektrisch betrieben so gut wie nichts an Geräusch produziert, zum anderen aber war das Entwicklungsunternehmen StreetScooter, eine Ausgründung der RWTH Aachen, schlicht so gut wie unbekannt. Es eilte den Professoren und Ingenieuren aus Aachen noch kein Ruf voraus, mit dem sie sich und ihrem StreetScooter-Projekt in Bonn hätten Gehör verschaffen können.

Am 15. September 2011 besuchte Bundeskanzlerin Angela Merkel die IAA in Frankfurt und schaute auch auf dem Stand der StreetScooter GmbH vorbei. Dort war das erste voll funktionstüchtige Elektrofahrzeug gleichen Namens zu bewundern. Von Professor Achim Kampker ließ sich die Kanzlerin alles erklären. Frau Merkel hörte und sah, was sie erfreute: bezahlbare Mobilität, neue Produktionstechnologie, Innovationsbeschleuniger, wirtschaftlicher Leichtbau. Nach Frankfurt war für den StreetScooter und für die Mannen um Prof. Kampker nichts mehr wie vorher. Denn nachdem Kanzlerin Merkel in den kleinen, knuffigen Stadtflitzer eingestiegen war, stieg schon bald die Deutsche Post AG ein – und das gleich in das ganze Unternehmen. Aber eins nach dem anderen.

Große deutsche und internationale Tageszeitungen berichteten von der IAA und von einer

At Deutsche Post in Bonn, nobody had ever heard of a company called StreetScooter until those days in September 2011. This may, on the one hand, have been because the electrically operated StreetScooter made next to no sound, and on the other hand, the development company StreetScooter, a spin-off of the RWTH-Aachen University, was virtually unknown. At that time the professors and engineers from Aachen didn't have any reputation to make their voices heard and get their StreetScooter project noticed in Bonn.

On September 15, 2011, German Chancellor Angela Merkel visited the IAA in Frankfurt and also dropped by the StreetScooter GmbH booth. There, the first fully functional electric vehicle of the same name was proudly on display. The Chancellor let Professor Achim Kampker explain it all to her. Mrs. Merkel heard and saw something that delighted her: affordable mobility, new production technology, an innovation accelerator, and cost-effective lightweight. After Frankfurt, nothing was ever the same for the StreetScooter and for the guys around Prof. Kampker. Because after Chancellor Merkel had gotten into the small, cuddly city car, the German Post AG soon got on board too, and they did the same for the entire company. But first things first.

Major German and international newspapers reported on the IAA and about a hitherto unknown company that had obviously devised something completely new for electromobility, one that above

Die erste Generation StreetScooter fährt bis zu 80 km/h schnell und ist primär für die Zustellung auf dem Land sowie in kleineren und mittleren Städten ausgelegt. Sie verfügt über eine Leistung von bis zu 30 kW, die von einer Lithium-Ionen-Batterie und einem Asynchronmotor erzeugt wird.

The first generation StreetScooter quickly goes up to 50 mph (80 km/h), and is designed primarily for delivery in rural areas and small and medium-sized cities. It has an output of up to 30 kW, which is generated by a lithium-ion battery and an AC induction motor.

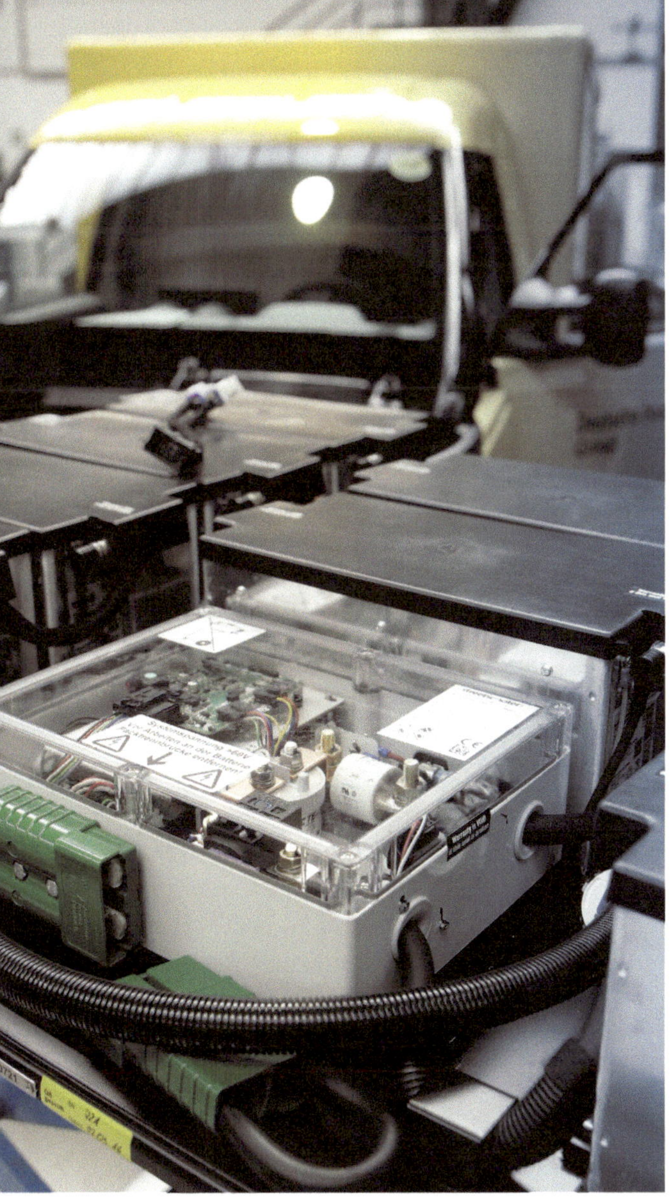

Ein Elektrofahrzeug wie der StreetScooter eignet sich insbesondere für Fahrten mit ausgeprägtem Start-Stopp-Verkehr. Durch die fast vollständige Emissionsfreiheit von Schadstoffen und Lärm sind die Autos sehr umweltfreundlich und leise.
An electric vehicle like the StreetScooter is particularly suitable for trips with heavy start and stop traffic. Thanks to their almost complete lack of emissions of pollutants and noise, the cars are extremely environmentally friendly and quiet.

bis dahin unbekannten Firma, die offensichtlich etwas völlig Neues in Sachen Elektromobilität gedacht hatte, vor allem aber radikal an den Entwicklungsprozess eines zukunftsfähigen Fahrzeuges herangegangen war. Als Jürgen Gerdes, Mitglied des Vorstands der Deutschen Post AG, einen der Artikel zum Thema las, klingelte es bei ihm – symbolisch gesprochen. Der DHL-Vorstand in Bonn erfuhr von einer Firma in Aachen und was Prof. Kampker der Zeitung zur Entwicklung des StreetScooters berichtete: „Entstehen sollte nicht das schnellste oder das leichteste Elektroauto, sondern eines, das bezahlbar ist und sich nicht erst bei sechsstelligen Stückzahlen zu bauen lohnt. Ein Fahrzeug, das ganzheitlich konzipiert ist und sich auf die Anforderungen der Praxis individualisieren lässt." Das kam einem Fahrzeug, wie Jürgen Gerdes es schon lange suchte, sehr nahe. Jetzt klingelte es in Aachen bei Prof. Kampker, und zwar das Telefon. „Nachdem sich DHL in Aachen gemeldet hatte, war ich begeistert", so Kampker. „Wir erkannten damals eine große Chance, mal tatsächlich etwas auf der Welt zu verändern. Dafür bin ich extrem dankbar."

all radically approached the development process for a sustainable vehicle. As Jürgen Gerdes, Board Member of Deutsche Post AG, was reading an article on the subject, he heard a bell go off – metaphorically speaking. The DHL Executive Board in Bonn learned about a company in Aachen and what Prof. Kampker reported to the newspaper about the development of the StreetScooter. "The goal isn't the creation of the fastest or the lightest electric car, but of one that is affordable and doesn't need to be produced in six-figure volumes to be cost-effective. A vehicle that is designed holistically and can be customized to meet the needs of its actual use." This sounded very much like a vehicle Jürgen Gerdes had been looking for some time. Now the "bell" that went off was in Aachen for Prof. Kampker – on his phone. "After DHL contacted us in Aachen, I was thrilled," says Kampker. "We realized at that time a great opportunity, not just for us, but for actually changing something in the world. For this, I am extremely grateful."

StreetScooter GmbH received an initial development contract for a vehicle, which had to be completely tailored to the requirements of

„AUTO-KONZERNE BLAMIERT: DEUTSCHE POST BESTELLT E-AUTOS BEI MITTELSTÄNDLER."[48]

"AUTO COMPANIES HUMILIATED: DEUTSCHE POST ORDERS E-CARS FROM SMES."[48]

Die StreetScooter GmbH erhielt einen ersten Entwicklungsauftrag für ein Fahrzeug, das ganz auf die Anforderungen der Zustellung und der Zusteller maßgeschneidert sein sollte. Bevor es zu dieser glücklichen Fügung zwischen der StreetScooter GmbH und der Deutschen Post AG kam, waren die DHL-Manager bei der deutschen Autoindustrie nicht fündig geworden. Ein funktionales und preiswertes E-Zustellfahrzeug, ohne Design-Schnickschnack, wollten oder konnten die etablierten Konzerne anscheinend nicht.

Das Team, das den DHL-Auftrag abwickeln sollte, konnte Prof. Kampker selbst bestimmen. Zusammen mit Ingenieurskollegen und Studenten begann Kampker jenes Elektrofahrzeug zu entwickeln, das die Post deutschlandweit für die leise und emissionsfreie Brief- und Paketzustellung nutzen wollte. Die Post hatte bereits klare Vorstellungen, wie ein Fahrzeug für die Zustellung optimal beschaffen sein müsste. Zudem wirkten knapp 150 DHL-Zusteller an der Entwicklung des elektromobilen Zustellfahrzeugs mit. Bereits 2012 konnte DHL das Nutzfahrzeug „Work" präsentieren.

In der Zustellung läuft der StreetScooter heute als einsitziges Post-Kompaktfahrzeug. Anstelle des zweiten Sitzes befindet sich neben dem Fahrersitz eine gelbe Zustellbox mit den Postsendungen, die jeweils als Nächstes auszuliefern sind. Der „Work" wurde eigens auf die Anforderung der Brief- und Paketzustellung zugeschnitten. 4,6 Meter lang, hat er einen Kastenaufbau, der genug Platz bietet für Brief- und Paket-Sendungen. Er verfügt zudem über eine robuste Ausstattung für den täglichen Einsatz unter den harten Bedingungen des Stop-and-go.

delivery and of the delivery person. Before this happy coincidence between the StreetScooter GmbH and Deutsche Post AG came about, the DHL managers had had no luck with the German car industry. The established companies seemed either unwilling or unable to provide a functional and inexpensive e-delivery vehicle without design frills.

Prof. Kampker was able to handpick the team to undertake the DHL job. Along with engineering colleagues and students, Kampker began to develop that electric vehicle that Deutsche Post would use across Germany for quiet and emission-free mail and parcel delivery. The Deutsche Post already had clear ideas about how a vehicle would have to be designed optimally for delivery. In addition, nearly 150 DHL deliverers were involved in the development of the electromobility delivery vehicle. By 2012, DHL was able to present the commercial vehicle "Work".

In the delivery business today, StreetScooter runs as a single-seat Deutsche Post compact vehicle. Next to the driver's seat instead of a second seat, there is a yellow delivery box with the mail and packages that need to be delivered next. The "Work" was specially tailored to the needs of letter and parcel delivery. At roughly 15 feet long (4.6 meters), it has a box structure, which provides enough space for mail and parcel shipments. It is also robustly built for everyday use under demanding stop-and-go conditions.

48 Quelle: Deutsche Wirtschafts Nachrichten, „Auto-Konzerne blamiert: Deutsche Post bestellt E-Autos bei Mittelständler", 14.03.21013 Source: Deutsche Wirtschafts Nachrichten, "Auto-Konzerne blamiert: Deutsche Post bestellt E-Autos bei Mittelständler", 14.03.21013

2014: DIE POST STEIGT EIN: KAUF DER STREETSCOOTER GMBH

2014: THE POST GETS ON BOARD: PURCHASES STREETSCOOTER GMBH

Heute baut die Deutsche Post ihre Autos für ihre Zusteller selbst. Das Know-how kommt von der RWTH Aachen. 2016 rollten bereits die ersten 2.000 Fahrzeuge in Aachen von der Produktionsstraße in den harten Einsatz. Nach und nach will die Post bis zu 30.000 Fahrzeuge durch den Streetscooter ersetzen. Mittelfristig plant man in Bonn sogar, die Fahrzeuge an Dritte zu vermarkten. So würde die umweltfreundliche StreetScooter-Idee auch in andere Bereiche des gewerblichen Alltags getragen. Post-Vorstand Jürgen Gerdes: „Wir betreiben eine der größten Fahrzeugflotten in Deutschland, darum haben wir ein besonderes Interesse, wirtschaftliche und emissionsarme Fahrzeuge einzusetzen. Das Kerngeschäft der Post ist über 500 Jahre alt – und zu jeder Zeit haben wir es mithilfe der modernsten verfügbaren Technik betrieben. Dass wir mit unserem eigenen StreetScooter auch heute wieder Innovationsführer in der Branche sind, passt ins Bild."[49]

Today the Deutsche Post builds its cars for its deliverers itself. The know-how comes from the RWTH Aachen. By 2016 the first 2,000 vehicles rolled off the production lines in Aachen right into heavy use. Gradually, Deutsche Post is looking to replace up to 30,000 vehicles with the StreetScooter. In the medium term, Deutsche Post is even planning to sell the vehicles to third parties. Thus, the environmentally friendly StreetScooter idea would be adopted in other areas of everyday commercial usage. Deutsche Post Board Member Jürgen Gerdes: "We operate one of the largest vehicle fleets in Germany, so we have a special interest in using economic and low-emission vehicles. The core business of the Deutsche Post is over 500 years old – and at all times we have run it using the most advanced technology available. So, the fact that with our own StreetScooter we are once again an innovation leader in the industry is no surprise."[49]

49 Quelle: Deutsche Wirtschafts Nachrichten, „Auto-Konzerne blamiert: Deutsche Post bestellt E-Autos bei Mittelständler", 14.03.21013
Source: Deutsche Wirtschafts Nachrichten, "Auto-Konzerne blamiert: Deutsche Post bestellt E-Autos bei Mittelständler", 14.03.21013

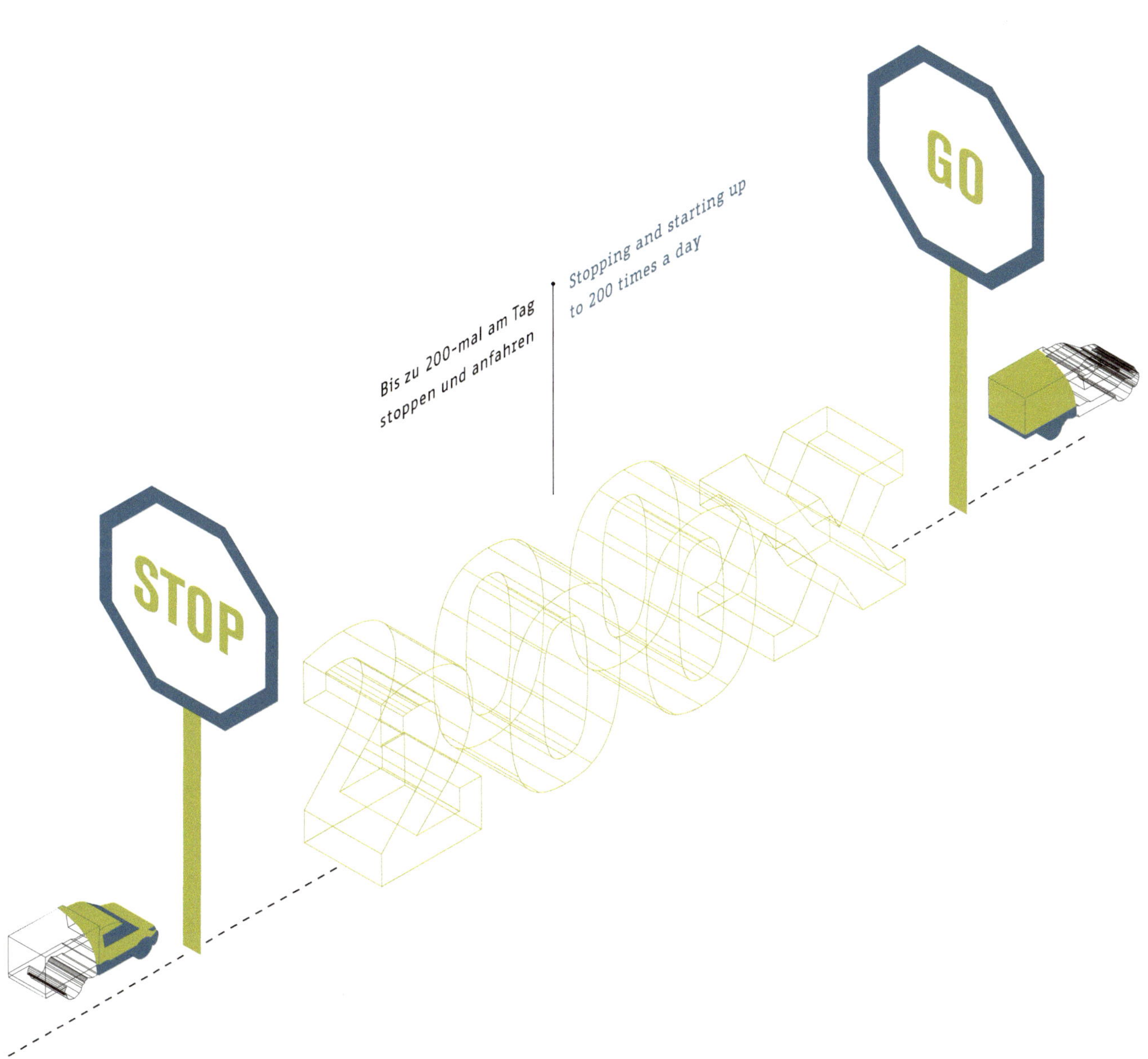
STOP
GO
Bis zu 200-mal am Tag
stoppen und anfahren
Stopping and starting up
to 200 times a day

„WENN WIR ÜBER UMWELTSCHUTZ REDEN, DANN MACHEN WIR AUCH UMWELTSCHUTZ UND KLEBEN NICHT NUR IRGENDWO IRGENDWELCHE STICKER DRAUF."

"IF WE'RE TALKING ABOUT ENVIRONMENTAL PROTECTION, IT'S BECAUSE WE'RE ALSO DOING ENVIRONMENTAL PROTECTION AND NOT JUST PUTTING A GREEN STICKER ON THINGS."

Jürgen Gerdes
Vorstand Deutsche Post, Ressort Post, eCommerce, Parcel
Chairman Deutsche Post, Ressort Post, eCommerce, Parcel

Warum glauben Sie an den StreetScooter?

Jürgen Gerdes: Sehen Sie, es gibt eine Frage, die wir uns immer wieder stellen: Wie können wir uns vom Wettbewerb differenzieren, auch in puncto Nachhaltigkeit, unseren Kunden deutlich machen, dass wir in allen Bereichen die bessere Wahl sind? Ich nenne das gerne den komparativen Konkurrenzvorteil. Und der zweite Fragenkomplex, damit im Zusammenhang, lautet: Glauben wir, dass es mehr eCommerce, mehr Pakete und damit mehr Logistik geben wird? Ja, das glauben wir, es wird mehr eCommerce geben, der unser Leben einfacher machen wird. Und ja, eCommerce führt zu mehr Paketen und damit zu mehr Verkehr. Aber wenn wir es richtig machen, führt mehr Logistik nicht zu mehr Umweltbelastung. Vor allem dann nicht, wenn man die letzte Meile bis zur Zustellung mit batteriebetriebenen Fahrzeugen bedient. Allerdings: Wenn man sich die Feinstaubbelastungen in China – oder nehmen wir nur allein eine Stadt wie Stuttgart – anschaut, sehe ich persönlich das Batterieauto noch lange nicht als der Weisheit letzten Schluss. Ich bin überzeugt: Es wäre nicht falsch, in Deutschland wie auch international das ganze Energiethema komplett neu zu durchdenken. Aber was wir als Unternehmen jetzt schon tun können, das tun wir auch – nicht zuletzt mithilfe des StreetScooters. Das setzt klare Zeichen in Sachen Umweltschutz und hebt uns zugleich vom Wettbewerb ab.

Wo ist Ihnen der StreetScooter zum ersten Mal über den Weg gefahren?

Ich verfolge das Thema E-Mobilität schon lange sehr genau. Denn mir war seit Jahren klar: Gerade für die Brief- und Paketzustellung eignen sich E-Mobile ideal. Nur, bei allem Respekt vor der Automobilindustrie: Auf dem Markt gab es leider keine passenden Fahrzeuge, und wir selbst sind keine Autobauer. Also sagte man mir: Dann geht es eben nicht. Nun bin ich allerdings ein Typ, der so etwas schwer akzeptiert und automatisch fragt: Was ist, wenn es doch geht? Dann las ich in einem Artikel von Professor Kampker und seinen Plänen.

Why do you believe in the StreetScooter?

Jürgen Gerdes: Well, there's this question that we ask ourselves again and again: How can we differentiate ourselves from the competition, also in terms of sustainability, making it clear to our customers that we are a better choice in all areas? I like to call it the comparative competitive advantage. And the second series of questions in this context is: Do we believe that there will be more eCommerce, more packages, and therefore more logistics? Yes, we believe there will be more eCommerce, which will make our lives easier. And yes, eCommerce leads to more packages and thus to more traffic. But if we do it right, more logistics won't lead to more pollution. Especially not when the last mile to the delivery is run with battery-powered vehicles. However: When you look at the particulate pollution in China – or even a city such as Stuttgart – I personally don't see the battery car as the be-all and end-all, and haven't for some time. I am convinced that it would be no mistake to completely rethink the whole energy issue, both in Germany and internationally. But what we can do as a company right now, we are doing – not least with the help of the StreetScooter. It makes a clear statement in terms of environmental protection and at the same time sets us apart from the competition.

Where did you first come across the StreetScooter?

I've been closely following the issue of electric mobility for a long time. Because for years I've known this: E-mobility is ideally suited for letter and parcel delivery. But, with all due respect to the automotive industry, there were no appropriate vehicles on the market, and we're not in the business of making cars. So, people told me: Well then, it won't work. But I'm really not the type of person who easily accepts something that, and I automatically asked. But what if it did? Then I read in an article by Professor Kampker and his plans. I found that immediately exciting. And I said to myself, if there's a company in Aachen that wants to build a small electric car, and I need a large

Das fand ich sofort spannend. Und ich habe mir gesagt: Wenn es in Aachen eine Firma gibt, die ein kleines E-Auto bauen will, und ich brauche ein großes E-Zustellauto – dann sollten wir doch mal versuchen, das zusammenzubringen. Das war der Startschuss für unseren ersten StreetScooter. Und siehe da: Auf einmal ging es doch. Dann haben wir eine sehr zügige Erfolgsgeschichte über 13 Monate hingelegt, bis das erste Auto lief und alltagstauglich war. Heute ist der StreetScooter ein Teil unserer Flotte, der nicht mehr wegzudenken ist. Weil er einfach genau in die Zeit passt und perfekt funktioniert.

Was waren denn die Hürden in der Entwicklung, über die Sie drüber mussten?

Wir haben ein Versuchsauto im Maßstab eins zu eins gebaut, ein sogenanntes „Lab Car". Ein richtiges Auto, an dem wir 150 Zustellerinnen und Zusteller haben Maß nehmen lassen: Große, Kleine, Dicke, Dünne, Junge, Alte, Männer, Frauen. Das ist extrem wichtig, wenn man bedenkt, dass unsere Zusteller zweihundert- bis dreihundertmal am Tag ein- und aussteigen müssen. Da muss die Ergonomie einfach stimmen, weil die Arbeit sonst im wahrsten Sinne auf die Knochen geht. Darum haben wir sehr viel Wert darauf gelegt, ein Fahrzeug zu bauen, das genauso ist, wie wir es im täglichen harten Einsatz brauchen. Aber eben auch nicht anders.

Einen rollenden Maßanzug?

Für mich ist mit dem StreetScooter etwas entstanden, was von der Ergonomie her alles bisher Dagewesene in den Schatten stellt. Und auch die anderen Dinge, auf die wir Wert gelegt haben, sind bemerkenswert: dass die Kosten im Rahmen bleiben, dass das Fahrzeug einfach zu reparieren ist, dass es einfach zu warten ist, dass es hochmodular ist.

Wo ist die Nachhaltigkeit an diesem Punkt?

Also, ich glaube nicht daran, dass man Autos ein paar Jahre lang abschreiben und sie dann

e-delivery car – then we should go ahead and have a go at joining forces to make it happen. That was the start of our first StreetScooter.

And lo and behold: all at once it actually worked out. Then we'd made a good start on a very rapid success story over 13 months, until the first car ran and was suitable for everyday use. Now the StreetScooter is a part of our fleet, which today is absolutely indispensable. Because it's exactly the right thing for our times and works perfectly.

So, what were the hurdles you had to leap over during the development process?

We built a test car at a scale of one to one, a "lab car". A real car. Then we let 150 postal carriers, big, small, fat, thin, young, old, men, and women test the ergonomics of it. This is extremely important when you consider that our deliverers have to get on and off two hundred to three hundred times a day. The ergonomics simply have to be right, because otherwise it's literally bone-tiring work. That's why we placed so much emphasis on building a vehicle that is just as we need it in daily, heavy use. It really couldn't be any other way.

A tailored suit on wheels?

For me something emerged with the Street-Scooter in terms of the ergonomics that eclipsed everything that came before. And then too, the other things that were important to us were remarkable: that the costs be kept in check, that the vehicle be easy to repair, that it be easy to maintain, that it be highly modular.

Where is the sustainability at this point?

Well, I do not believe that you write off cars for a few years and then have to exchange them, just because you think, now's the time for it. Instead, I believe that to some extent these things can be used for much longer and that this type of resource conservation is a very important factor when considering the issue of sustainability in a holistic way. If the seat is damaged, we build a

austauschen muss, nur weil man eben meint, jetzt sei die Zeit dafür gekommen. Sondern ich glaube, dass man diese Dinge teilweise viel länger nutzen kann und dass diese Art von Ressourcenschonung ein ganz wesentlicher Faktor ist, wenn man das Thema Nachhaltigkeit einmal gesamthaft betrachtet. Wenn der Sitz defekt ist, bauen wir einen neuen Sitz ein. Aber wenn der Rahmen noch X Jahre länger hält, dann läuft das Fahrzeug mit diesem Rahmen eben noch X Jahre weiter. Auch das ist Nachhaltigkeit, erzielt durch Ressourcenschonung.

Die Deutsche Post ist jetzt im Prinzip Automobilbauer. Was kann sie denn besser als die Automobilindustrie?

Ich denke nicht, dass wir Automobilbauer sind. Wir sind vielmehr Werkzeugbauer. Ich betrachte den StreetScooter in erster Linie als Werkzeug. Er leistet uns perfekte Dienste bei der Erfüllung unserer Aufgaben. Und viele emotionale Dinge spielen – anders als beim klassischen Automobil – eine untergeordnete Rolle. Wir haben versucht, maximal Rücksicht auf unsere Anforderungen zu nehmen und genau das herzustellen, was wir brauchen.

Stichwort Customer Experience und Customer Obsession.

Ja, dieses Fahrzeug wurde genau nach unseren Kriterien als spezieller Verwender mit ganz speziellem Profil gebaut. Das ist Customer Experience pur.

Das ist eine spannende Frage, was Customer Experience im Automobilbereich überhaupt bedeutet. Denn eigentlich wird sie zu sehr ausgeblendet, oder?

Ich bin mir nicht sicher, ob schon jeder verstanden hat, wie wichtig das ist. Und in meinen Augen sind die aktuellen Megatrends wie Digitalisierung und Automatisierung eine Riesenchance, um auch in dieser Branche unterschiedliche Zielgruppen noch besser zu bedienen. Private wie gewerbliche. Viele Unternehmen und viele, die

new seat. But if the frame holds for another X years longer, then the vehicle runs on this frame for another X years. That is also sustainability, achieved through resource conservation.

In principle, Deutsche Post is now an automaker. What can they actually do better than the automotive industry?

I don't think that we're automakers. We're more like toolmakers. I primarily think of the StreetScooter as a tool. It provides us with perfect service in fulfilling our tasks. And unlike the classic automobile, many emotional things play a subordinate role. We've tried to give maximum consideration for our needs and to produce exactly what we need.

Keywords: customer experience and customer obsession.

Yes, this vehicle was built exactly according to our criteria as a specific user with a very specific profile. That's pure customer experience.

That's an interesting question, what does customer experience even mean in the automotive industry? Because they've become much too far removed, wouldn't you say?

I'm not sure if everyone's truly understood how important that is. And my view is that the current megatrends such as digitization and automation provide a tremendous opportunity to better serve different target groups in this industry. Both private and commercial. Many companies and many who work for them would gladly share their requirements and their experience with the automotive industry. That's been the case with me personally for a long time, too.

What do you think is fundamentally standing in the way of a new thinking or new developments?

There are always legitimate and understandable factors, that cannot be dismissed out of hand. If I employ thousands of people in engine production, if I have invested and am investing

dort Verantwortung tragen, würden sich zu ihren Anforderungen und ihren Erfahrungen gern mit der Automobilbranche austauschen. Auch auf mich persönlich trifft das seit Langem zu.

Was, glauben Sie, steht grundsätzlich einem neuen Denken oder Neuentwicklungen im Weg?

Es gibt immer ganz berechtigte und nachvollziehbare Faktoren, die man nicht von der Hand weisen kann. Wenn ich tausende von Menschen in der Motorenproduktion beschäftige, wenn ich Millionen, vielleicht sogar Milliarden in Motorenentwicklung investiert habe und investiere, dann kann ich nicht einfach von heute auf morgen Produktion und Entwicklung komplett auf Elektroantrieb umstellen. Es geht ja auch um traditionelle Kernkompetenz. Aber das Rad dreht sich trotzdem immer weiter. Auch Kernkompetenz ist immer auf einen gewissen Zeitraum begrenzt. Und wenn es eine technologische Neuerung gibt, dann kann das in der Tat dazu führen, dass eine Kernkompetenz überdacht und womöglich neu justiert werden muss. Und der Druck von außen ist ja da.

Ich war vergangene Woche in Indien. Da waren es um die 35 Grad. Aber die Sonne hat man trotzdem nicht gesehen, und blauen Himmel gab es auch nicht. Wenn wir jetzt nach China fahren würden, nach Peking, Shanghai oder wohin auch immer, wäre es ähnlich. Umweltverschmutzung ist kein Randthema mehr. Umweltverschmutzung ist ein zentrales Thema. Wir haben nur diese eine Erde. Und deswegen muss uns allen eines wichtig sein: der Versuch, die Geschäftstätigkeit auch auf Umweltbelange auszurichten.

Um den Faden mal weiterzuspinnen: Wo sehen Sie denn Märkte für dieses exemplarische Beispiel StreetScooter? Für das Werkzeug, wie Sie sagen?

Ich kenne viele Unternehmen und ganze Branchen, in denen der StreetScooter zum Einsatz kommen könnte, weil es dort ein ähnliches Bewegungsprofil gibt. Was den StreetScooter für eine Reihe von Zielgruppen attraktiv macht, ist ja seine hohe Variabilität, vor allem beim Aufbau. Bei uns

millions, perhaps even billions in engine development, then I can't just change production and development entirely to electric drive overnight. It's also about traditional core competence. But the wheel of time keeps turning. Even core competence is always limited to a certain time period. And if there is a technological innovation, that can actually lead to the need to rethink a core competence and where possible, adjust it. And the pressure from outside is already there.

I was in India last week. It was about 95 degrees there. And yet you didn't see the sun and there were no blue skies either. If we were to now go to China, to Beijing, Shanghai or wherever, it would be much the same. Pollution is no longer a fringe issue. Pollution is a key issue. We only have one Earth. And that's why one thing needs to be important to us all: the effort of aligning economic activity to environmental concerns.

To follow along these lines a bit: Where do you see markets for this exemplary example, StreetScooter? For the tool, as you call it?

I know many companies and whole industries in which the StreetScooter could be used because they've got a similar movement profile. What makes the StreetScooter attractive for a number of target groups is indeed its high degree of variability, especially in its construction. In our operation there are now already three different vehicle sizes in use. In addition, we will soon see a model that drives much faster than the previous ones and for much longer distances. For the purposes of Deutsche Post and DHL Paket, that is not needed, but I still find it very impressive to observe how the technology can evolve. Other possibilities will very quickly emerge.

So, is the plan for the whole fleet to be electrified in the long term?

Wherever possible, we will electrify the majority of our fleet, yes. Because it just makes sense.

im Betrieb sind inzwischen schon drei verschiedene Fahrzeuggrößen im Einsatz. Außerdem werden wir demnächst ein Modell sehen, das deutlich schneller fährt als die bisherigen und deutlich länger. Für die Zwecke von Post und DHL Paket ist das nicht erforderlich, aber ich finde es schon auch sehr beeindruckend wahrzunehmen, wie sich die Technologie weiterentwickeln lässt. Da werden sich schnell weitere Einsatzmöglichkeiten zeigen.

Soll denn langfristig die ganze Flotte elektrifiziert werden?

Wo immer möglich, werden wir den größten Teil unserer Flotte elektrifizieren, ja. Weil es einfach Sinn macht.

Da wird immer diese Geschichte von den Elektrofahrzeugen der Post vor dem Zweiten Weltkrieg kolportiert …

Ja, die gab es.

Dann ist die Post wahrlich ein Pionier der Elektromobilität?

Anfang des zwanzigsten Jahrhunderts war die Post in dieser Hinsicht in der Tat ganz weit vorn. Irgendwann wurde dann aber der Verbrennungsmotor Antrieb der Wahl, auch für uns.

Wie man jetzt sieht, können bestimmte Dinge wiederkommen, wenn die Zeit dafür reif ist. Aber dann vielleicht doch auch in ganz anderer Form und vor einem völlig unterschiedlichen Hintergrund.

Wird der StreetScooter immer ein gewerbliches Thema bleiben? Ein Werkzeug?

Das muss nicht unbedingt so sein. Wir haben bestimmt noch die eine oder andere Idee, von der ich glaube, dass sie durchaus spannend sein könnte.

Die Nachhaltigkeit wird weiter an Bedeutung gewinnen?

Auf jeden Fall! Nicht nur, was unsere Branche und unser Geschäft angeht. Aber zweifellos

There's been this rumor about the electric vehicles of the Deutsche Post that's been going round since before World War II …

Yes, there was.

Then the Deutsche Post is truly a pioneer of electric mobility?

Early in the twentieth century the Deutsche Post was actually quite advanced in this respect. But then at some point the combustion engine drive became the drive of choice, even for us.

As you can see now, certain things can come back when the time is right for them. But then perhaps also in a completely different form and against a completely different background.

Will the StreetScooter always remain an industrial topic? A tool?

This needn't necessarily be so. We certainly still have one or two other ideas that I think could be quite exciting.

Will sustainability become increasingly important?

Definitely! Not only in terms of our industry and our business. But undoubtedly our ability to truly deliver CO_2 neutrality in conurbations is a very significant issue. We are the absolute leader over the competition in this.

Germany in 20 years. How do you foresee mobility on the road ahead?

I foresee some trends that will have a few implications for Germany. Our mobility will be quite different than it is today! Digitalization will also bring with it great changes and ensure that we are much smarter, more compact and space saving on the road. There will be driverless cars, we will latch into systems that holistically manage and organize traffic and thereby ensure for greater safety. In order to be able to provide mobility on the road in increasingly growing agglomerations, there will be additional concepts that go beyond a purely road traffic context and extend to logistics.

ist schon einmal unsere Fähigkeit, wirklich CO_2-neutral in Ballungsräumen zuzustellen, ein sehr erhebliches Thema. Damit stehen wir im Wettbewerb absolut führend da.

Deutschland in 20 Jahren. Wie stellen Sie sich die Mobilität auf der Straße vor?

Ich sehe da einige Trends voraus, die eines zur Folge haben werden: Unsere Mobilität wird ganz anders sein als heute! Die Digitalisierung wird auch hier noch große Veränderungen mit sich bringen und dafür sorgen, dass wir viel smarter, kompakter und platzsparender unterwegs sind. Es wird autonom gefahren, wir werden uns in Systeme einklinken, die den Verkehr ganzheitlich steuern und organisieren und damit idealerweise auch für größere Sicherheit sorgen. Um in weiter stark wachsenden Ballungsräumen überhaupt noch Mobilität auf der Straße gewährleisten zu können, wird es zusätzlich Konzepte geben, die über den rein straßengebundenen Verkehr hinausgehen und sich auf die Logistik erstrecken. Ich denke da an Lufttransporte, die ja heute schon ernsthaft getestet werden; aber ebenso an eine viel intensivere Nutzung unterirdischer Möglichkeiten für die Ver- und Entsorgung. Also ein sehr vielschichtiges Thema mit ganz unterschiedlichen Facetten. Wovon ich aber insgesamt überzeugt bin: Die Mobilität wird irgendwann hoch performant sein im Sinne von Customer Experience und gleichzeitig sehr, sehr umweltschonend und sehr, sehr effizient.

I am thinking of air transportation, which is already being seriously tested; but also of a much more intensive use of underground facilities for supply and disposal. So, it's a very complex issue with diverse facets. What I'm convinced of overall is this: Mobility will eventually be high performing in terms of customer experience and also very, very environmentally friendly and very, very efficient.

STREETSCOOTER IN ZAHLEN UND FAKTEN

STREETSCOOTER IN FACTS AND FIGURES

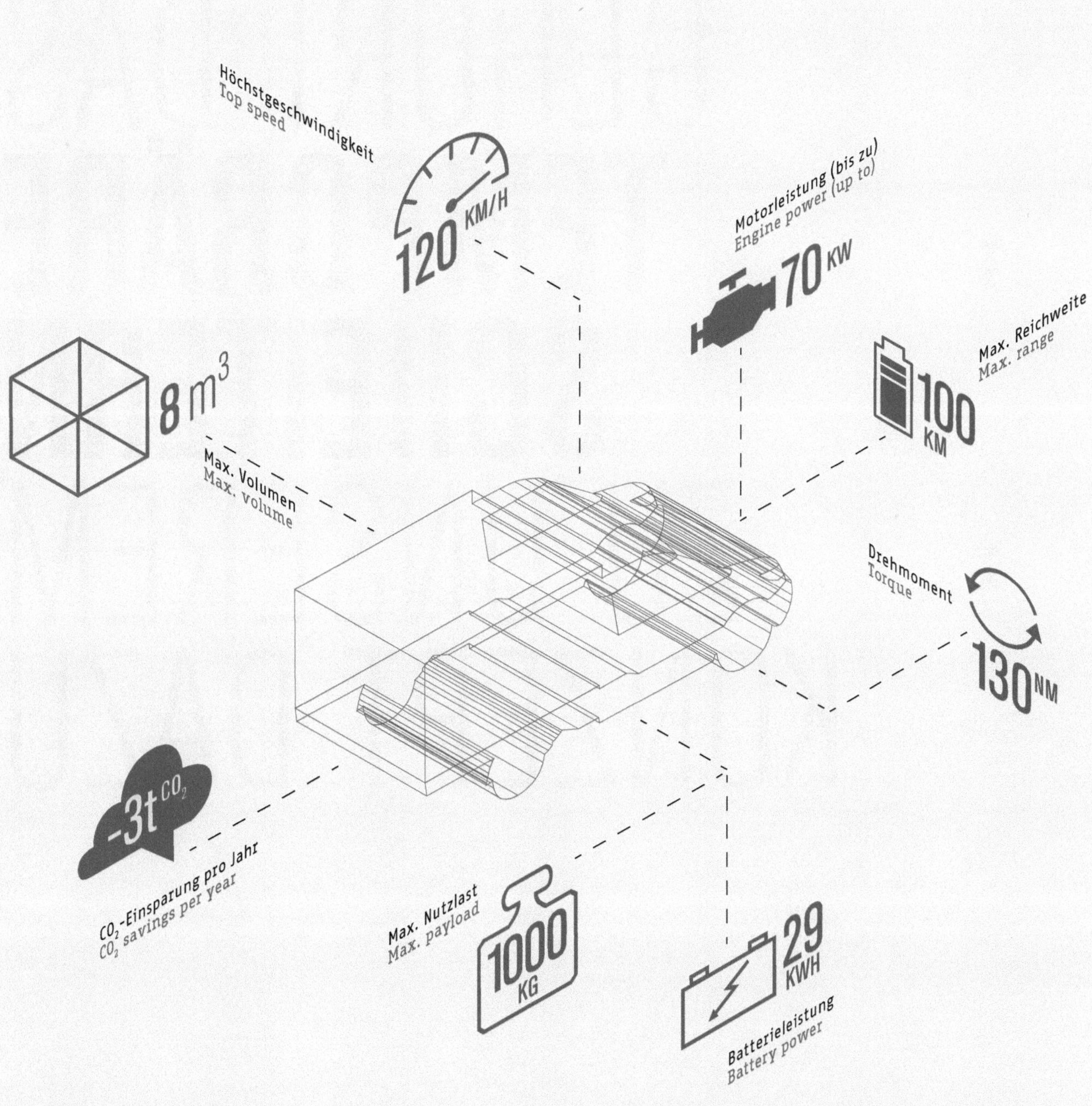

„ NATÜRLICH GEHÖRT DAS **TEMPORÄRE SCHEITERN** ZU UNSEREM KONZEPT DAZU.

TEMPORARY FAILURE IS CERTAINLY PART OF OUR APPROACH. BUT LOSING IS DEFINITELY NOT.

Prof. Achim Kampker
Geschäftsführer der Streetscooter GmbH
Executive Vice President E-Mobilität Deutsche Post DHL
CEO of Streetscooter GmbH
Executive Vice President E-Mobility Deutsche Post DHL

ABER GANZ SICHER NICHT DAS VERLIEREN.

10

DIE STREETSCOOTER-MOBILITÄTSLÖSUNG

THE STREETSCOOTER MOBILITY SOLUTION

Effizient elektrisch durch die Stadt mit dem „Work", „E-Bike" oder „E-Trike". Die StreetScooter GmbH bietet die passende Lösung für den Flotteneinsatz. Auf zwei, drei und vier Rädern. Hundert Prozent elektrisch, hundert Prozent CO_2-frei.

Efficient. Getting around the city electrically with the "Work", "E-Bike" or "E-Trike". StreetScooter GmbH offers the right solution for fleet use. On two, three, and four wheels. One hundred percent electric, one hundred percent CO_2-free.

ELEKTROMOBILITÄT WEITERGEDACHT

THE NEXT STEP FOR ELECTROMOBILITY

„Weiter so, Herr Kampker." Mit diesen Worten beendete Bundeskanzlerin Angela Merkel im September 2011 ihren Pflichtbesuch auf dem Stand der StreetScooter GmbH. Gedanklich war Professor Achim Kampker sicher schon auf dem Weg in eine Zukunft, in der Ökonomie und Ökologie weiter verschmelzen sollten. In jedem Fall müsste die Zukunft in seinen Augen stromlinienförmig sein, stromlinienförmig im bildlichen Sinne. Das Weltbild des StreetScooter-Teams unter seiner Führung ist eindeutig ohne sauberen Strom, sprich Elektromobilität, nicht zu denken. Und mit dem „Compact", wie das Ur-Modell des StreetScooters hieß, war Ende 2011 ein Anfang gemacht. Von da an sollte es mit einer geballten Ladung an Motivation und Ideen weitergehen. Anders als viele Ingenieurskollegen glaubte Professor Kampker nicht an das „Immer höher, schneller und weiter". Ihn trieb und treibt etwas anderes an. „Ich glaube nicht, dass das die richtige Philosophie ist. Ich denke, dass Ingenieure sich fragen sollten, wie kann ich Technik dafür nutzen, dass die Welt lebenswert ist und erhalten bleibt".[50]

Vier Beispiele, wie das geht, zeigt die aktuelle Modellreihe der StreetScooter GmbH. Für die Zustellung im urbanen Umfeld bietet die Aachener Firma mit dem „Work", dem „Work L" und den Zwei- und Dreirad-E-Bikes eine perfekte Kombination aus Ökonomie und Ökologie.

"Keep up the good work, Mr. Kampker". These were the words Chancellor Angela Merkel left Professor Kampker with in September 2011 during her stop at the StreetScooter GmbH booth. Conceptually, Professor Achim Kampker was certainly already well on the way to a future in which economy and ecology would be fused even further. In his view, the future would in any case be both literally and figuratively streamlined. The worldview of the StreetScooter team under his leadership is clearly inconceivable without clean power, i. e. electromobility. And with the "Compact", as the original model of StreetScooter was called, he'd already taken a big first step in late 2011. From then on, it would take off, fueled by impressive quantities of motivation and ideas. Unlike many of his engineering colleagues, Professor Kampker did not subscribe to the "Ever Faster, Higher, Farther" approach. He was and is driven by something different. "I don't think this is the right philosophy. I think that engineers should be asking, How can I use technology to ensure that the world is better place to live in and can be preserved."[50]

The current line of models from StreetScooter GmbH provides four examples of just how that works. For delivery in the urban environment, the Aachen company offers a perfect combination of economy and ecology with its "Work", "Work L" as well as the two- and three-wheeled e-bikes.

50 Quelle: Deutsche Verkehrs-Zeitung (DVZ), „Achim Kampker und der StreetScooter", 30.01.2017
Source: Deutsche Verkehrs-Zeitung (DVZ), "Achim Kampker und der StreetScooter", 01/30/2017

Außen gelb, innen grün: Modell „Work"

Der „Work" ist die intelligente Antwort auf alle Fragen bezüglich Kosteneffizienz, Ergonomie und Nachhaltigkeit. Zur besonderen Wirtschaftlichkeit dieses Modells tragen die reduzierten Wartungs- und Servicekosten (um etwa 50 %) und die reduzierten Reparaturkosten (um bis zu 80 %) bei sowie der geringe Stromverbrauch, die wirtschaftliche Attraktivität des Stroms im Vergleich zu Benzin bzw. Diesel und der Verzicht auf einen Großteil mechanischer Verschleißteile. Nachhaltig beim „Work" sind nicht nur seine Antriebstechnologie, sondern auch neue Entwicklungs- und Produktionsprozesse, die Individualität und Effizienz gleichermaßen ermöglichen. Die Stärken des „Work" basieren unter anderem auf einer kundenorientierten, ergonomischen Gestaltung, einem dem Nutzungsprofil adaptierten Fahr- und Batteriemanagementsystem für bestmögliche Reichweiten sowie einer robusten und somit sicheren und wartungsarmen Bauweise. Diese ermöglicht einen strapazierfähigen Einsatz sowie die leichte Austauschbarkeit von Karosserie- und Fahrzeugteilen.

Der „Work" wurde für die Logistikanforderungen im urbanen Umfeld entwickelt. Im Mittelpunkt steht ein umfassendes Individualisierungskonzept. Die verschiedenen Aufbauten erlauben eine maßgeschneiderte Konfiguration. Großabnehmern

Yellow on the outside, green on the inside: the "Work" model

The "Work" is the intelligent answer to any questions about cost efficiency, ergonomics and sustainability. Contributing to the specific economics of this model are the reduced maintenance and service cost (about 50 %) and the reduced repair cost (up to 80 %) plus the low power consumption, the economic attractiveness of electricity compared to gasoline or diesel, and the waiving to a great extent of mechanical wear parts. When it comes to sustainability in the "Work", it's not just about its drive technology; it's also new development and production processes, which allow for individuality and efficiency. The strengths of the "Work" are based in part on a customer-oriented ergonomic design, a usage profile adapted to the driving and battery management system for optimum cruising range and a robust and therefore safe and low-maintenance design. This allows for a durable use and easy replacement of body and vehicle parts.

The "Work" was specifically developed to meet the logistics requirements in urban areas, with a focus on a comprehensive individualization concept. The different assembled parts make it possible to easily customize the configuration. For large-scale buyers, the "Work" even

bietet der „Work" sogar eine Entwicklung und Produktion von spezifischen Komponenten auf individuelle Praxis-Anforderungen hin an.

offers a development and production of specific components based on their specific usage requirements.

Mehr dran, mehr drin: Modell „Work L"

Der größere Bruder des „Work" heißt „Work L", d.h. mehr Lade-Volumen und damit mehr Zuladungsmöglichkeit. Durch seine modulare Bauweise ist der kleinere „Work" einfach skalierbar. Zudem verfügt der „Work L" über einen stärkeren Elektromotor (45 kW Dauerleistung/70 kW Spitzenleistung) und eine leistungsstärkere Lithium-Ionen-Batterie mit einer Reichweite von bis zu 100 Kilometern. Dem Basismodell des „Work" stehen 38 kW Motorleistung (48 kW Spitzenleistung) und eine Reichweite von 80 km zur Verfügung. Das Maximaltempo des StreetScooters in der „L"-Version liegt bei abgeregelten 80 km/h; offen sind bis zu 130 km/h drin. Geschwindigkeit und Motorleistung sind auf die Anforderungen der Deutschen Post ausgelegt und können kundenindividuell angepasst werden, sodass auch höhere Reichweiten möglich sind.

More to it, more in it: the "Work L" model

The big brother of the "Work" is called "Work L", where the "L" stands for more Load volume and thus more Load capacity. The modular design of the smaller "Work" makes it easily scalable. The "Work L" is also equipped with a stronger electric engine (45 kW continuous power/70 kW peak power) and a more powerful lithium-ion battery, giving it a range of up to 60 miles (100 kilometers). The basic model of the "Work" has 38 kW of engine power (48 kW peak power) and a range of 50 miles (80 km). Speed and engine power are designed to meet the requirements of Deutsche Post and can be customized to meet individual customer requirements.

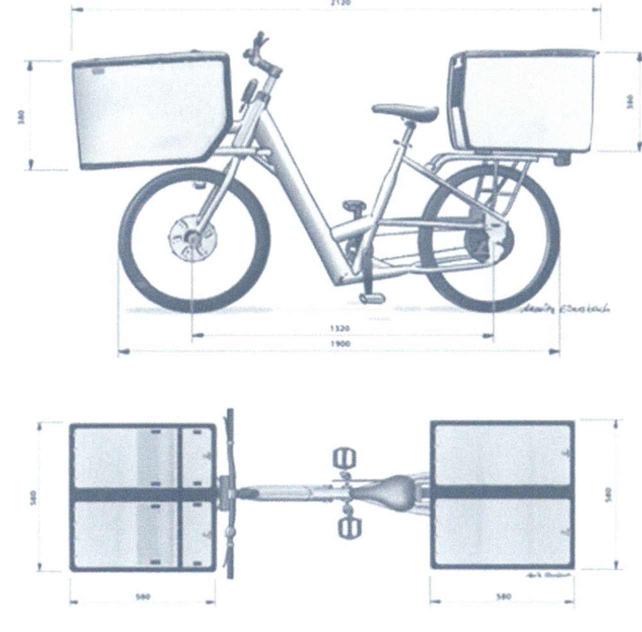

Frischzellen-Tour: Modell „E-Bike"

Das E-Bike von StreetScooter verbindet dynami-schen Fahrspaß, Geschwindigkeit, Wendigkeit und hohe Nutzlast. Der einzigartige Aufbau und der elektrische Antrieb setzen Maßstäbe bei der Pro-duktivität. Die Effizienz beginnt bei der Ergonomie für den Fahrer. Das E-Bike bietet einen einzigar-tigen Sitzkomfort. Das ausgewogene Fahrverhal-ten und der agile elektrische Antrieb bieten neben dem Fahrkomfort auch hohe Nutzlastmöglichkei-ten. Das „E-Bike" von StreetScooter wurde mit zusätzlichem Stauvolumen ausgerüstet und bietet Nutzlast bis zu 50 kg. Mit dem zukunftsgerichteten elektrischen Antrieb wird öffentlichkeitswirksam ein Zeichen für ein besseres Stadtklima gesetzt.

Elektrisch aufgeladen: Modell „E-Trike"

Ohne Schadstoffe und mit voller Kraft durch den engen Stadtverkehr. Das StreetScooter „E-Trike" ist ein echtes Kraftpaket: einfaches Handling, hohe Zuladung und eine robuste Konstruktion für

Tour de Freshness: the "E-Bike" model

The E-Bike from StreetScooter combines dynamic driving pleasure, speed, maneuverability and high payload. The unique design and the electric drive set new standards for productivity. Its efficiency starts with the ergonomics for the rider. The E-Bike offers seat comfort second to none. In addition to its driving comfort, the balanced handling and agile electric drive also provide high payload options. The "E-Bike" from StreetScooter was equipped with additional storage volume and offers payloads of up to 110 lbs. (50 kg). With its forward-looking electric drive, it effectively sends a clear and pub-lic signal for a better urban climate.

Electrically charged: the "E-trike" model

Getting through tight city traffic with real power and zero emissions. The StreetScooter "E-Trike" is a real powerhouse: easy handling, high load capa-city, and a robust design for everyday use. The E-Trike from StreetScooter offers solid roadholding,

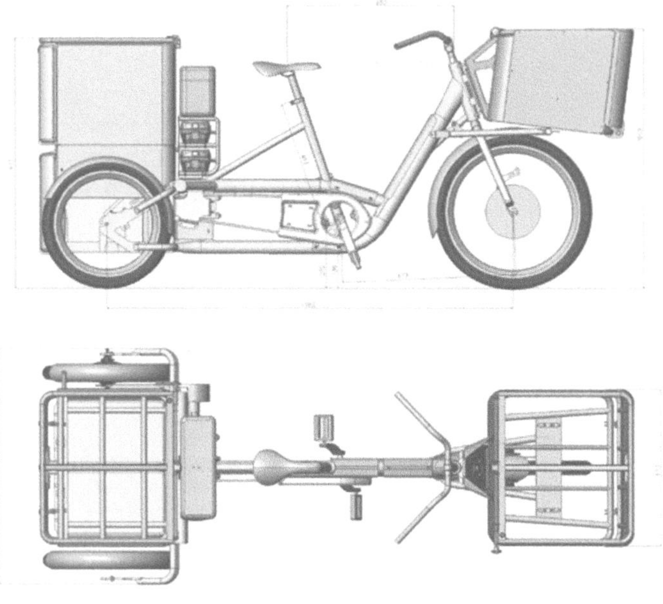

den täglichen Einsatz. Das E-Trike von StreetScooter bietet solide Bodenhaftung und beruhigende Sicherheit mit beeindruckender Wendigkeit. Bei der Verbindung mit dem elektrischen Antrieb setzt StreetScooter neue Maßstäbe bei der Produktivität. Die hohe Kurvenstabilität und der agile elektrische Antrieb bieten viel Fahrspaß und zugleich hohe Nutzlastmöglichkeiten. Das E-Trike wurde mit zusätzlichem Stauvolumen ausgerüstet und bietet Platz für 6 Briefbehälter mit bis zu 90 kg Nutzlast.

Für die Kurzstrecke, aber nicht kurz gedacht!

Der StreetScooter ist nicht nur ein E-Fahrzeug, das ganz für die Kurzstrecke gebaut ist, er transportiert auch einen ganzheitlichen Systemgedanken, der auf Nachhaltigkeit ausgelegt ist. Diesem Gedanken folgend, hat die StreetScooter GmbH neben den verschiedenen Fahrzeugmodellen zahlreiche Lösungen entwickelt, die alle auf eine Reduktion der Flottenkosten ausgerichtet sind. Dazu zählt zum Beispiel das Lademanagement.

and reassuring safety with impressive agility. With the connection to the electric drive, StreetScooter sets new standards in productivity. The high cornering stability and agile electric drive provide loads of driving pleasure and at the same time high payload options. The "E-Bike" was equipped with additional storage volume and offers room for 6 letter containers with a payload of up to 200 lbs. (90 kg).

Made for the short haul – but not the short-term!

The StreetScooter isn't just an electric vehicle designed for the short haul: it also conveys a holistic system approach designed for sustainability. Following this line of thinking, StreetScooter GmbH has developed numerous solutions in addition to the various vehicle models, all of which are aimed at reducing fleet costs. For instance, charge management. In practice this means that if in the evenings, the Deutsche Post delivery vehicles more or less simultaneously drove back to their depots, and

Praktisch bedeutet das: Wenn die Zustell-Fahrzeuge der Deutschen Post abends mehr oder weniger gleichzeitig auf ihre Betriebshöfe fahren und dann an die Ladeschnittstellen angeschlossen werden, steigt der Strombedarf und damit der Strompreis. Damit das nicht passiert, hat StreetScooter intelligente Steuerungssysteme entwickelt, die den Ladevorgang der Fahrzeuge laufend justieren und so Peaks in der Last vermeiden.

Eine weitere Flottenmanagement-Technologie zur Kostenoptimierung ist „Car-to-Cloud". Die StreetScooter-Ingenieure haben dazu eine transparente und echtzeitfähige Car-to-Cloud-Box entwickelt, die permanent eine Aufbereitung der Fahrzeugdaten ermöglicht. Ziel ist es, den Status der Fahrzeuge aus dem laufenden Betrieb heraus analysierbar zu machen, damit jederzeit Daten für eine eventuelle Anpassung und Weiterentwicklung vorliegen.

Natürlich ist auch die Batterie ein ganz großes Thema im Rahmen der ökologischen und ökonomischen Nachhaltigkeit. StreetScooter bietet Kunden nicht nur Lösungen zur Weiternutzung alter Batterien in stationären Speichern an, sondern auch ihr in Fahrzeugflotten erprobtes Batterie-Management-System (BMS), dazu Module und Energiespeicher-Systemlösungen inklusive Individualisierung und Beratung – und auf Wunsch natürlich auch Support und Service.

are then connected to the charging interfaces, electricity demand goes up, driving the cost of electricity up with it. To prevent this from happening, StreetScooter has developed intelligent control systems that continually adjust the charging of the vehicles, avoiding load peaks.

Another fleet management technology to optimize costs is "car-to-cloud". The StreetScooter engineers have developed a transparent and real-time car-to-cloud box that allows permanent processing of the vehicle data. Its aim is to make the status of the vehicles in operation analyzable so that data for a possible adaptation and further development is available at any time.

The battery is obviously another major issue in the context of environmental and economic sustainability. StreetScooter not only offers customers solutions for further use of old batteries in stationary storage systems, but also their vehicle fleets proven battery management system (BMS), plus modules and energy storage system solutions including personalization and consulting – and, on request, of course, support and service.

Prof. Achim Kampker
Geschäftsführer der Streetscooter GmbH
Executive Vice President E-Mobilität Deutsche Post DHL
CEO of Streetscooter GmbH
Executive Vice President E-Mobility Deutsche Post DHL

„ALS WIR AN DAS PROJEKT STREETSCOOTER RANGINGEN, WAREN WIR

EINES GANZ SICHER NICHT: MACHBARKEITS-ILLUSIONISTEN.

AS WE REALLY GOT DOWN TO WORK ON THE STREETSCOOTER PROJECT, THERE WAS ONE THING WE DEFINITELY WEREN'T: FEASIBILITY ILLUSIONISTS.

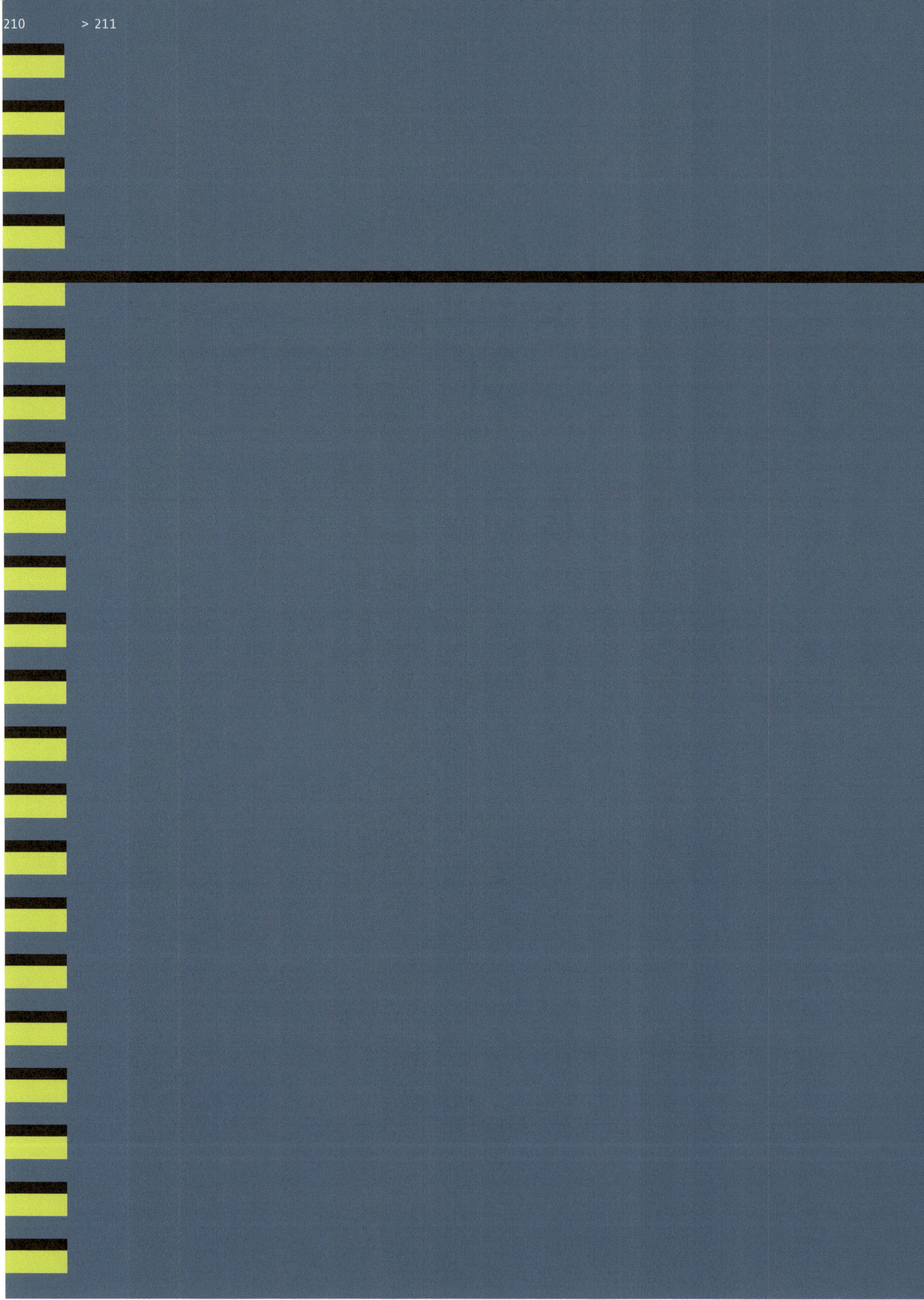

11

DIE STREETSCOOTER-VISION

THE STREETSCOOTER VISION

Märchen beginnen mit „Es war einmal", Erfolgsgeschichten mit „Es wird einmal". Aus dem Primotyp des StreetScooters wurde ein Zustellfahrzeug, praktisch, ökologisch, günstig. Wie geht es weiter? Die Deutsche Post plant eine grüne Zukunft und der StreetScooter ist fester Bestandteil dieses Plans.

Fairy tales begin with "Once upon a time"; but success stories begin with "There will be a time". From the StreetScooter primotype, a delivery vehicle was born that was practical and environmentally friendly. What's next? Deutsche Post is planning a green future and StreetScooter is an integral part of that plan.

VOM SHORT-DISTANCE-VEHICLE ZUM NACHHALTIGKEITS-TREIBER

FROM SHORT-DISTANCE VEHICLE TO SUSTAINABILITY DRIVER

Die Deutsche Post verfolgt ehrgeizige Klimaziele. Bis 2020 will sie ihre CO_2-Effizienz um 30 % verbessern. Das Unternehmen setzt dabei auf Knowhow aus Aachen, wenn es um die Zukunft der Elektromobilität geht. Nach und nach kommen immer mehr Elektroautos vom Typ StreetScooter zum Einsatz. Während sich der StreetScooter bundesweit öffentlichkeitswirksam durch die Straßen der Städte und Regionen bewegt, ist man in Aachen schon einen weiteren Schritt in Richtung Zukunft unterwegs. Und zwar mit dem fahrerlosen StreetScooter. Der autonom fahrende StreetScooter soll dem Postzusteller nicht die Arbeit wegnehmen, sondern dessen Job erleichtern. Wie das? Der StreetScooter im Modus autonomous folgt den Zustellern beim Austragen der Pakete und Briefe in Schrittgeschwindigkeit. Die Idee: Die Zusteller müssen die Postautos nicht ständig wieder vorfahren. Die Technik funktioniert bereits, nur die Erlaubnis für den Praxisversuch steht noch aus.

Deutsche Post is pursuing ambitious climate targets. By 2020, it wants to improve its CO_2 efficiency by 30 %. And when it comes to the future of electromobility, the company relies on know-how from Aachen. Gradually, more and more StreetScooter-type electric cars are being put into operation. While the StreetScooter is on the road across the cities and regions of Germany, making a strong statement about Deutsche Post's commitment, back in Aachen, they're already busy taking the next step into the future – with the driverless StreetScooter. The self-driving StreetScooter isn't meant to take jobs away from mailmen; it's meant to make their jobs easier. But how? When in self-driving mode, the StreetScooter follows the delivery people at walking pace as they deliver packages and letters. The idea? The delivery people won't have to constantly drive the mail car up, inching house-by-house anymore. The technology already works; the guys in Aachen are just waiting for the green light to run field testing.

Autonomes Fahren und andere E-Volutionen

Autonomous driving and other e-Volutions

Die Tests laufen auf der Avantis-Versuchsstrecke. Die Deutsche Post DHL baut dazu den Avantis-Innovationspark in Aachen/Heerlen an der deutsch-niederländischen Grenze zum Post-Testgelände für die Entwicklung innovativer Zustellformen auf und

The tests are run on the Avantis test track. Deutsche Post DHL is setting up and expanding the Avantis Innovation Park in Aachen/Heerlen on the German-Dutch border at the Deutsche Post test site for the development of innovative delivery methods.

Garrelt Duin
Minister für Wirtschaft, Energie,
Industrie, Mittelstand und Handwerk
des Landes Nordrhein-Westfalen
*Minister of Economy, Energy,
Industry, Small Business and Crafts
of North Rhine-Westphalia*

„Aus einer Idee wird ein erfolgreiches Produkt geboren – ein Top-Beispiel für Wissenstransfer. Der StreetScooter ist ein Meilenstein der Elektromobilität, aber seine Wirkung strahlt weit darüber hinaus: Innovation hat eine Heimat in NRW."

"From an idea, a successful product is born – a prime example of knowledge transfer. The StreetScooter is a milestone of electric mobility, but its effect extends far beyond that: Innovation has a home in North Rhine-Westphalia."

aus. Das grenzüberschreitende Gelände umfasst eine Fläche von rund 20.000 qm. „Unser Unternehmen will seine CO_2-Effizienz bis zum Jahr 2020 um 30 % verbessern, verglichen zum Basisjahr 2007. Im Rahmen unseres Pilotprojekts, CO_2-freie Zustellung in Bonn, setzen wir auf den Einsatz von Elektrofahrzeugen. Aufgrund der guten Erfahrungen möchten wir unsere Aktivitäten im Bereich innovativer Zustellformen verstärken"[51], so Jürgen Gerdes, Konzernvorstand für den Unternehmensbereich Post – eCommerce – Parcel. „Der Testbetrieb von speziell für die Brief- und Paketzustellung entwickelten Elektrofahrzeugen soll nun am neuen Standort intensiviert und ausgebaut werden, um dieses Ziel zu erreichen."[51]

Auf dem Gelände arbeitet ein Prüfzentrum zum Testbetrieb von Elektrofahrzeugen. Außerdem werden weitere innovative Zustellkonzepte entwickelt und getestet. Parallel wird die Forschungsarbeit vorangetrieben. Visionär geht es dabei immer auch darum, die Erfolgsstory des StreetScooters

The cross-border site covers an area of about 215,000 square feet (20,000 square meters). "Our company wants to improve its CO_2 efficiency by 30% by 2020, compared to the base year 2007. As part of our pilot project, CO_2-free delivery in Bonn, we are relying on the use of electric vehicles. Based on the good experience we've had, we would like to strengthen our activities in innovative delivery methods,"[51] says Jürgen Gerdes, Group Chairman for the Post – eCommerce – Parcel division. "The test operation of electric vehicles specially designed for mail and parcel delivery will now be intensified at the new location and expanded in order to achieve this goal."[51]

There is a test center on the site working on test operation of electric vehicles, and other innovative delivery solutions are also being developed and tested. Research is moving ahead in parallel. It's also about keeping a visionary approach, so that the next chapter to the StreetScooter success story can be written. "With our location in Aachen,

51 Quelle: AVANTIS GOB NV, „Deutsche Post DHL baut Innovationspark auf Avantis", www.avantis.org, 2014
Source: AVANTIS GOB NV, "Deutsche Post DHL baut Innovationspark auf Avantis" [Deutsche Post DHL building innovation park at Avantis], www.avantis.org, 2014

weiter voranzutreiben. „Mit unserem Standort in Aachen sind wir bestens aufgestellt, um den Testbetrieb neuer Mobilitätsarten und -wege auf höchstem Niveau voranzutreiben, insbesondere aufgrund der Nähe zur exzellenten Hochschullandschaft. Die gemeinsam mit der RWTH Aachen entwickelten Elektrofahrzeuge werden bereits in verschiedenen Zustellstützpunkten im Alltagsbetrieb dem Praxistest unterzogen und sollen nun auch in die Weiterentwicklung der Verbundzustellung einbezogen werden, bei der Brief- und Paketzustellung zugunsten der Effizienz gemeinsam erfolgen“[52], betont Gerdes.

Darüber hinaus beteiligt sich die StreetScooter GmbH gemeinsam mit universitären und industriellen Partnern an Förderprojekten, die die Forschung und Entwicklung zukunftsweisender Technologien beinhalten. Offene Fragestellungen im Bereich der Elektromobilität werden von StreetScooter adressiert sowie innovative Projektideen in interdisziplinären Konsortien realisiert, um damit einen Beitrag zur bundesweiten Etablierung von elektrischen Fahrzeugflotten zu leisten.

Forsch in die Zukunft: ein E-Flitzer für die Stadt

„Mit diesem hippen Flitzer, der unglaublich viel Spaß macht, werden wir auf dem Markt für Elektrofahrzeuge einen Nerv treffen“[53], sagt Professor

we are ideally positioned to drive test operation of new mobility modes and routes forward at the highest level, especially thanks to the proximity to its excellent universities. The electric vehicles developed in collaboration with RWTH Aachen are already subjected to practice testing in everyday use in various delivery depots and will now also be included in the further development of joint delivery, where mail and parcel delivery are done together for optimal efficiency,"[52] says Gerdes.

StreetScooter GmbH is also participating jointly with academic and industrial partners to support projects that include research and development of forward-looking technologies. StreetScooter addresses questions in the field of electric vehicles and implements innovative project ideas in interdisciplinary consortia so it can contribute to the nationwide establishment of electric vehicle fleets.

Dashing into the future: An e-speedster for the city

"We'll be making a real splash in the market for electric vehicles with this hip, incredibly fun speedster,"[53] says Professor Günther Schuh, referring to the next big thing from the concepts and development shop of RWTH Aachen. Behind all this is an innovative short-range vehicle that is aimed at a young, urban target group that's interested in innovative technology. That's precisely why e.GO

52 Quelle: AVANTIS GOB NV, „Deutsche Post DHL baut Innovationspark auf Avantis", www.avantis.org, 2014
Source: AVANTIS GOB NV, "Deutsche Post DHL baut Innovationspark auf Avantis" [Deutsche Post DHL building innovation park at Avantis], www.avantis.org, 2014

53 Quelle: DIE WELT, „Deutsche Post plant grüne Zukunft", www.welt.de, 24.03.2015
Source: DIE WELT, "Deutsche Post plant grüne Zukunft" [Deutsche Post plans green future], www.welt.de, 03/24/2015

Nachhaltige Mobilität für urbanes Leben:
der e.GO Life (hier ein Prototyp)
Sustainable mobility for urban life:
The e.GO Life (here a prototype)

Günther Schuh und meint damit das nächste big thing aus der Ideen- und Entwicklungsschmiede der RWTH Aachen. Dahinter steckt ein innovatives Kurzstreckenfahrzeug, das sich an eine junge, urbane Zielgruppe wendet, die Interesse an innovativer Technologie hat. Dazu wurde eigens die e.GO Mobile AG ins Leben gerufen, am 1. April 2015 wurde die Gesellschaft ins Handelsregister eingetragen und hat mit diesem Datum auch ihr operatives Geschäft begonnen. Die Firma e.GO plant einen besonders leichten Zweisitzer. Als Variante ist ein e.GO mit höherer Antriebsleistung und sportlicher Performance geplant. Prof. Schuh möchte das größere Modell mit mehreren E-Motoren ausstatten, die spontan 150 Pferdestärken an Spitzenleistung auf die Straße bringen. „Wenn man die Variante mit vier Motoren nimmt, kann man auch jeden Porsche an der Ampel stehen lassen"[53], so Prof. Schuh. Und er denkt weiter: Er sieht auch neue Möglichkeiten für den Verkauf seiner Fahrzeuge. Das Internet bietet hier ungeahnte Wege.

Mobile AG was created, which was entered into the commercial register on April 1, 2015, starting its operational business on that very day. The e.GO company is planning a particularly lightweight two-seater. They're also planning an e.Go variant with higher drive power and a more sporty performance. Prof. Schuh would like to equip the larger model with several e-engines that could instantly give it 150 horsepower at peak performance when it hits the road. "If you buy the variant with four engines, you'll also be able to accelerate so fast you'll leave any Porsche standing at the light[53]," says Professor Schuh. Furthermore, he also sees new opportunities for the sale of its companion, the StreetScooter. The Internet offers unprecedented options here.

"And with the help of licensees, distribution into markets such as the US, Mexico and China is perfectly possible,"[54] says Schuh with a view to the future. Right, the future: "Such a city car like the e.GO lasts as long as a streetcar does – so about

54 Quelle: DIE WELT, „Deutsche Post plant grüne Zukunft", www.welt.de, 24.03.2015
Source: DIE WELT, "Deutsche Post plant grüne Zukunft" [Deutsche Post plans green future], www.welt.de, 03/24/2015

„Und mit Hilfe von Lizenznehmern ist auch ein Vertrieb auf Märkten wie den USA, Mexiko und China vorstellbar"[54], so Schuh mit Blick in die Zukunft. Ach ja Zukunft: „So ein Stadtauto wie der e.GO hält so lange wie eine Straßenbahn, also etwa 30 Jahre, mit dem Effekt, dass auch mehrfache Wiederverkäufe samt technischer und optischer Überholung beim Nutzerwechsel denkbar sind."[54]

E-Motion inklusive:
E-Kart mit Pedelec-Antrieb

Als dritte und besonders günstige Variante ist ein E-Kart in der Entwicklung, vergleichbar mit einem Kettcar mit Pedelec-Antrieb. Nach Meinung von Professor Schuh haben Elektromobile eben nicht nur mit Vernunft zu tun, sondern auch mit Emotionen. Als autobegeisterter Wissenschaftler und Unternehmer ist Schuh gerne auch mal mit seinem Sohn und zwei Prototypen des E-Kart in Aachen unterwegs.

30 years – which means that even multiple resales including technological and aesthetic overhaul when there's a change of users is conceivable."[54]

E-Motion included:
E-Kart with pedelec drive

As a third and particularly inexpensive option, there is an e-Kart in development, comparable to a Kettcar with a pedelec drive. As Professor Schuh sees it, electric cars aren't just about an appeal to reason, but to emotions too. As scientist and entrepreneur who's also a car enthusiast, Schuh sometimes also loves to tool around Aachen with two prototypes of the E-Kart, son in tow.

So bunt wie das Leben:
der e.GO Life
As colorful as life:
The e.GO Life
e GO Life

„MAN KANN DEN STREETSCOOTER IN VIER WORTEN ERKLÄREN: LANGFRISTIG. EINFACH. BESSER. MACHEN."

THE STREETSCOOTER CAN BE EXPLAINED IN FOUR STEPS: IN THE LONG RUN. SIMPLY. MAKE IT. BETTER.

Prof. Achim Kampker
Geschäftsführer der Streetscooter GmbH
Executive Vice President E-Mobilität Deutsche Post DHL
CEO of Streetscooter GmbH
Executive Vice President E-Mobility Deutsche Post DHL

LANGFRISTIG

1. Wettbewerbsvorsprung – dauerhaft:
Effizienzsteigerung im Industrialisierungsprozess führt zu einem langfristigen Wettbewerbsvorsprung.

EINFACH

2. Reduktion – radikal:
Zielsetzung ist eine radikale Reduktion des Industrialisierungsaufwands bei gleichzeitiger Erhöhung des Kundenwerts.

BESSER

3. Vorgehensweise – strukturiert:
9 Lösungsprinzipien detaillieren die Vorgehensweise zur Maximierung des Return on Engineering.

MACHEN

4. Umsetzung – erfolgreich:
Referenz: Nach nur 3 Jahren von der Idee bis zur Serienproduktion ist der StreetScooter heute im Einsatz bei der Deutschen Post.

IN THE LONG RUN

1. Competitive advantage - permanent:
Increased efficiency in the Industrialization process leads to a long-term competitive advantage.

SIMPLY

2. Reduction – radical:
The aim is a radical reduction of the industrialization effort while increasing customer value.

BETTER

3. Procedure – structured:
9 principles detailing the procedure for maximizing the Return on Engineering.

MAKE IT

4. Implementation – successful:
Reference: After only 3 years from idea to series production, the StreetScooter is in use today at Deutsche Post.

STREETSCOOTER:
den Industrialisierungsprozess neu erfunden

STREETSCOOTER:
the industrialization process reinvented